39

Fortschritte der chemischen Forschung
Topics in Current Chemistry

Computers
in Chemistry

Springer-Verlag
Berlin Heidelberg GmbH 1973

ISBN 978-3-540-06231-8   ISBN 978-3-540-38510-3 (eBook)
DOI 10.1007/978-3-540-38510-3

Library of Congress Catalog Card Number 51-5497.

Contents

Introduction

As in other branches of science and technology, so the use of computers in chemistry is steadily increasing. However, the expectations about what computers can do for the chemist are often rather unrealistic, sometimes too optimistic, less frequently too pessimistic, but in most cases rather subjective. These articles try to provide material for a realistic evaluation of the possibilities and limitations of computer applications in the field of chemistry.

What is a computer? Basically it is a machine which can process symbols. To use it as a tool for the solution of problems one has to:

a) identify all parameters describing the initial state of the given problem and describe their values in terms of symbols that can readily be processed by the computer

b) identify all parameters characterizing all possible final states of the given problem and describe all their possible values by symbols easily processed by the computer

c) formulate an algorithm to transform these symbols in such a way that they will represent a solution to the given problem.

Step c) corresponds to the writing of the computer program proper. Steps a) and b) may be called system analysis. In other words, they answer the questions:" What has to be done?" and "How is it to be done?" For most practical purposes system analysis (*i.e.* answering the question "What has to be done?") is a much more time-consuming and sophisticated task than writing the program proper.

A sufficient knowledge of one or more suitable programming languages and of basic mathematics is required for writing the program. Furthermore, besides a generous measure of common sense, the major prerequisites for the system analyst are a profound understanding of the problem to be solved, and broad perspective of the system. Thus, chemists at home in the problem complex have to become part of the team at an early stage. For efficient cooperation with the programmers, the chemist should have a working knowledge of computer technology and programming. Unfortunately, at most European universities these subjects are not taught at all or are treated rather academically.

Most programs available today are system-oriented solutions to special problems and, as such, very efficient for solving that problem but generally not easily adaptable to other needs. In fact, adaptibility and flexibility are commonly not taken sufficiently into account from the very beginning of system analysis. In the end, this leads to constant and costly reformulation of the same algorithm by different people at different times. Such duplication can be avoided by having the program written in a standardized form and generalized so as to provide a reasonable measure of adaptibility. Furthermore, potential users should be alerted to the availability of such programs in the specialist journals. This presents a problem in itself, as chemical journals usually reject detailed program descriptions as being non-chemical in nature, and experience teaches us that chemists do not read data-processing journals.

There is plenty of room for the use of computers in chemistry. Certain analytical methods, such as X-ray structure analysis, Fourier transform spectroscopy, and the investigation of complex mixtures with GLC/MS combinations, are unthinkable without direct access to a powerful computer. Likewise many branches of theoretical chemistry (*e.g.* MO calculations) need large computing facilities. Chemical technology benefits from the fast optimization of chemical processes by simulation. The control of complete production lines by computers has become almost routine. The rapid and selective dissemination and processing of the chemical information which is available in the chemical literature should help to eliminate the costly duplication of work in research and development. The computer-assisted planning of synthesis is just beginning. This list of examples could be further extended at will.

Systems utilizing one large computer are generally better than systems using several small ones for different purposes. On the other hand, a large computer system represents a sizable investment, so it is often advantageous to build up a system in steps, thereby distributing the investment over a longer time span. At the same time, the organizational and human problems involved in trying to run several complex real-time programs on the same computer are virtually absent when the same tasks are individually assigned to smaller computers. The decision as to whether the system of choice is to be one large computer or several smaller ones has to be made for every case with due consideration of all the circumstances. Ready-made decisions concerning the use of computers in chemistry cannot be offered, as up to now research has been mainly concerned with just a few selected fields. The reason for this is most probably to be found in the chemist's rather under-developed computer conciousness. The trends in computer technology are surely increasing capacity at decreasing prices, so that this situation will certainly change for the better in the near future.

The Coming of the Computer Age to Organic Chemistry:

Recent Approaches to Systematic Synthesis Analysis

Ajit J. Thakkar

Department of Chemistry, Queen's University, Kingston, Ontario*, Canada, and Lehrstuhl für Theoretische Chemie der Technischen Universität München

Contents

---

* Present address

I. Introduction

The solution of a synthetic problem of organic chemistry involves at least three distinct, although not necessarily independent, steps. These include the formulation, in outline, of a synthetic strategy, the selection and ordering of specific individual steps, and the experimental execution of the synthesis.

The importance of a synthetic plan can hardly be overemphasized. As pointed out by R. B. Woodward [1]:

> "... synthetic objectives are seldom, if ever, taken by chance, nor will the most painstaking or inspired, purely observational activities suffice. Synthesis must always be carried out by plan, and the synthetic frontier can be defined only in terms of the degree to which realistic planning is possible, utilizing all of the intellectual and physical tools available."

It should be apparent then that an analysis of the general principles involved in the formulation of such a plan would be a step toward a deeper understanding of synthesis, and hence also a step forward in the evolution of the art of synthesis to higher forms.

Admittedly, the task of systematizing the procedure for developing a synthetic plan seems quite hopeless in view of the tremendous number of organic compounds known to exist, the fantastic diversity of reactions available for use in synthesis, and the uncertain and narrow limits on the scope of any particular reaction. However diffidence has never been the way of science and the path has already been paved for those who will follow the pioneers.

An investigation into the methodology of any scientific discipline usually can be separated into three or four distinct stages: (1) definition of the discipline and its goals, (2) articulation of the implicit assumptions or axioms upon which the discipline is based, (3) critical analysis of the methods employed by the practitioners of the discipline in light of the information obtained in the earlier stages of the investigation, and (4) the extension of an extant method, the development of a new method, or the indication of promising avenues for further development of the methodology of the discipline.

II. The Fundamental Assumptions of the Art of Synthesis

Since the first step, *i. e.* definition of the discipline and its goals, has already been dealt with in the introductory paragraphs, it is possible to proceed with the articulation of the implicit assumptions and axioms of synthesis.

Corey [2] has pointed out the following considerations that can be regarded as axiomatic to synthesis:

(1) The various elements involved in the solution of a synthetic problem are interdependent and hence inseparable. Although it may be necessary to separate them for the purposes of simplifying an analysis, compensation must be made for this factor at a later stage of the analysis.

(2) It is possible to generate an enormous number of possible synthetic routes to a complex molecule.

(3) These possible pathways can be derived by the recognition of structural units within a molecule that can be formed or assembled by known or conceivable chemical reactions. For convenience, these structural units are called "synthons".

(4) A synthetic route should start from readily available substances.

(5) There are definite criteria by which the relative merits of various projected synthetic routes can be evaluated. These criteria would serve to eliminate a large number of inferior alternatives and they include:

    (i) The probability of achieving the desired change at each step in the sequence should be high. It follows that competing reactions should be minimized and yields maximized.

    (ii) Potential alternative routes should exist particularly where one or more of the individual steps is dubious.

    (iii) The solution should be simple.

(6) The mechanism and scope of the reactions chosen should be reasonably well-understood.

III. Problem Solving Techniques Applicable to Synthesis

The third step of this investigation, a critical review of recent attempts at the analysis and development of synthetic methodologies, follows. Chemists employ a variety of problem-solving techniques of varying sophistication, depending upon both the chemist and the problem, to devise synthetic routes to a given compound. The extreme methodologies which differ with respect to analytical and logical sophistication and generality have been termed [3] the "direct-associative" and "logic-centred" approaches to synthesis.

The direct-associative approach is employed when the chemist directly recognizes within a structure a number of readily available and undisguised subunits which can be brought together in the proper way using standard reactions with which he is very familiar. For example, compound 3 can be assembled from acetone *1* and benzaldehyde *2* by two successive Claisen-Schmidt condensations followed by dehydration:

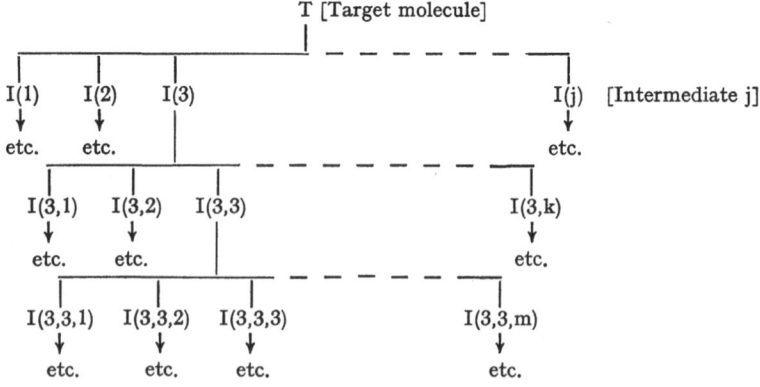

Most synthetic chemists would perceive this possible route to compound *3* with a minimum of logical analysis or planning simply because of the fact that the subunits are so obvious and so familiar, like the reactions required to join them, that the simple processes of mnemonic association lead directly to a possible solution. Obviously the direct-associative approach is limited to relatively simple synthetic problems.

The other extreme is the logic-centered approach which is limited only by the frontiers of chemistry and the potential of human intelligence and creativity. The molecule to be synthesized, the target molecule, is logically analyzed to generate a set of intermediates which can be converted in a single synthetic step to the target molecule. Iteration of this procedure for each intermediate so generated results in the development of a "synthetic tree" (Fig. 1). Iteration is continued until readily available starting materials are reached. Of course, the various synthetic routes resulting from the generation of the synthetic tree require further analysis in order that their relative merits be evaluated.

Fig. 1. The generation of a synthetic tree

In practice, most chemists use an approach that is neither direct-associative nor logic-centred but a combination of both. The chemist generates a partial tree of synthetic intermediates until he comes across an intermediate whose structure is simple enough for him to use the direct-associative approach.

It is clear that the logic-centred approach is superior to the direct-associative approach because the former does not make any assumptions as to starting materials. However the tremendous amount of time and effort required to generate and analyze a complete synthetic tree has resulted in the logic-centred approach being used very infrequently, if at all, by practising organic chemists.

IV. The Computer as a Tool for Synthesis Planning

The iterative nature of the procedure used to generate a synthetic tree suggests the possibility of using an electronic computer. Chemists who oppose the increasing use of computers in chemistry may do well to take note of a statement made by Professor Raymond Cattell [4]:

> *"In many fields of human endeavour, from war games to space travel and clinical psychology, we have perhaps to get used to the idea of helping out the limited memories which underlie our judgments by electronic aids, and when this is more widespread ....."*

A. LHASA: The Harvard Approach

Research into the use of digital computers to assist in the derivation of synthetic routes to complex molecules has been in progress for several years at Havard. The program is called LHASA (*Logic and Heuristics Applied to Synthetic Analysis*). The research is being carried out by a team headed by Corey.[3,5—9]

It will be useful to use a few terms coined by Corey and Wipke [3,5] in the discussion of this work. Since the logic-centred approach of generating a synthetic tree involves analytic processes which depend heavily upon the structural features of reaction products and the consideration of molecular changes in the retro-synthetic sense, the direction of computer analysis is termed 'antithetic' as opposed to the direction of laboratory execution which is termed 'synthetic'. A process in the antithetic direction is called a 'transform' while a process in the synthetic direction is called a 'reaction'. A reaction is represented by $\xrightarrow{(s)}$ while a transform is represented by $\xRightarrow{(a)}$. A brief summary of the key features of the computer program currently developed follows.

The first problem to be solved is that of *man-machine communication*. Corey, Wipke and others [3,6] have developed a system by which the chemist feeds in a structure by drawing a standard two-dimensional structural formula using an electrostatic tablet (Rand tablet) and a Rand pen. The computer communicates with the chemist by displaying conventional structural formulae on a cathode ray tube. Provision is also made for hard copy output using a plotter. As emphasized by Corey and Wipke [3,5] this allows for an interactive relationship between the chemist and the computer.

The second problem to be solved is that of *structural representation within the computer*. Corey, Wipke and others [3,5] have used a specific form of connectivity tables for both bonds and atoms to achieve this. The atom part of the table contains for each atom an arbitrary sequence number, the number of attachments, the charge, the valence and the type (C, S, O, etc.) as well as the relative address of the first bond entry for this atom. The coordinates of this atom for display on the cathode ray tube are also stored. The bond part of the table contains two entries for each bond between atoms. Each entry includes an arbitrary sequence number for the bond, the sequence numbers for the attached atoms, the bond type (single, double, etc.) and the relative address of the bond entry for the next attached atom. A sample table for cyclopropane is shown in Fig. 2.

ATNO	NATCH	CHARGE	VALENCE	TYPE	POINTER[a]
1	2	1	4	C	906
2	2	1	4	C	905
3	2	1	4	C	903

ATNO	BNDNO	BNDTYP	POINTER[b]	LOCATION
2	3	1	0	901
3	3	1	0	902
1	2	1	901	903
3	2	1	0	904
1	1	1	902	905
2	1	1	904	906

Fig. 2. A sample connectivity table for cyclopropane

---

[a] To first bond entry for this atom.
[b] To the bond entry for the next attached atom in number sequence.

In order to make maximum use of core storage[c], intermediates generated from the target molecule are internally represented in LHASA by the physical differences between the connectivity tables of the target structure and the intermediate.

The next problem to be solved is that of developing a method by which the computer can *gather and store synthetically significant structural information* for later use in the generation of a synthetic tree by the application of chemical knowledge. Corey, Wipke and others have developed a method [7] in which the key step is the creation of sets containing binary information as to which atoms or bonds possess a particular property. Some binary sets for acetone are illustrated in Fig. 3. The binary digits 0 and 1 indicate, respectively, that the corresponding atom or bond is not or is a member of the set in question. The leftmost digit corresponds to the first atom or bond according to sequence number, and each digit to the right corresponds to the next atom or bond in the sequence.

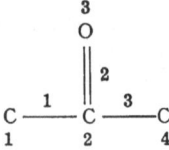

SETNAME	SET
CARBON	1101
OXYGEN	0010
BOND2SET (atoms)	0110

Illustration of set operations: Carbonyl oxygens: OXYGEN AND BOND2-SET

Fig. 3. Sample binary sets and set operations for acetone

Starting with the most basic sets, increasingly complex sets can be constructed by standard set operations. The latter can be achieved on the computer by using basic logical instructions such as AND or the inclusive OR. This is also exemplified in Fig. 3 where the intersection of the OXYGEN (set of all O atoms) and BOND2SET (atoms to which at least one double bond is attached) sets yields the set of all carbonyl oxygens.

---

[c] In order to illustrate the types of techniques that may be used to make maximum use of core storage, an algorithm for the efficient storage of connectivity tables is presented in Appendix I.

Sequences of such logical manipulations of sets are used to recognize functional groups. Al algorithm for the machine perception of synthetically significant rings has also been written [9]. Other types of structural information perceived by LHASA include features such as aromatic ring systems, relative levels of reactivity of each type of functional group, and asymmetric centres.

The next problem to be solved is that of developing a method by which the computer can use the structural information perceived in the previous step to *generate a synthetic tree* by the application of suitable transforms. Since the identification of applicable transforms is made on the basis of target structure and is independent of offspring structure, Corey *et al.* have used a method [8] that classifies transforms according to the nature of the critical structural features of a target molecule to which the transforms may be keyed. By way of illustration a few of the more important classes of transforms are shown below:

(1) Transforms which create two functional groups in the synthetic product (group pair transforms). *e.g.* aldol transform.

(2) Transforms which change one functional group and which also modify the structural skeleton in the synthetic product (single group transforms), *e.g.* the Grignard transform.

(3) Single group transforms which modify only a functional group (functional group interchange), *e.g.* oxidation transform.

(4) Ring transforms which form or modify some particular type of ring, *e.g.* the Diels-Alder transform.

The program also includes provision for the evaluation of possible transforms applicable to the target structure. The program consists of a table which stores chemical information about various classes of transforms, a transform-choosing module which matches the features of a target structure against the data table, evaluates and stores all applicable transforms, and a transform-executing module which executes the chosen transforms one-by-one to generate new intermediates in the synthetic tree.

The final, and perhaps the most fascinating, problem to be solved is that of *strategy selection* [5] in the course of the generation of a synthetic tree. A few of the possible types of stratigies enumerated by Corey include:

(1) Strategies based on particular structural characteristics of a target molecule. For example, the derivation of a synthetic route to ethyl N-benzoylhomomeroquininate [4], a key intermediate in the total synthesis of quinine [10], could be based upon the strategy of focussing upon the vinyl group.

4 Ethyl N-benzoylhomomeroquininate

(2) Strategies based on the selection of certain key transforms, or more generally, certain chemical information, the application of which becomes a goal. For example, the derivation of a synthetic route from compound *5* to compound *6*, a key intermediate in the total synthesis of reserpine [11], could be based upon the strategy of using an osmium tetroxide oxidation followed by a sodium periodate cleavage of the vicinal hydroxyl groups so generated.

11

$$5 \qquad 6$$

(3) Opportunistic strategies. These are usually found by chance. For an example, see the route from compound *5* to compound *6* actually chosen by Woodward et al. [11].

The Harvard program currently lacks stereochemical capability, and work is under way [5] to rectify this situation. Corey feels that the task of developing a general problem-solving procedure for use by a computer is too large to be accomplished in a five or ten-year period and is unlikely to be complete, in a final sense, in the forseeable future [5].

It is nevertheless clear that when the program is eventually completed and becomes readily available to all organic chemists, it will take its place among the indispensable tools a chemist routinely uses to assist him in his work. However, the shortcomings of the program are equally obvious. The empirical nature of the program, insofar that it uses empirical information about known chemical reactions, dictates its limitation to the generation of synthetic pathways which are suggested to the chemist by his knowledge of the literature. It should be borne in mind that the program is designed to function in a manner analogous to a chemist. A machine that mimics a human can only produce the same results, albeit in a very much shorter interval of time, as a human can. It is clear that science has a long way to go before it can even claim to be a satisfactory approximation to truth. Hence it follows that the limitations of the program will be the limitations of present-day science. In other words, the program will by no means present a closed solution to the problems of synthesis.

B. The Stony Brook Program

Another empirical program has been developed by Gelernter et al. [12], who undertook the project as a study in an application of artificial intelligence to a combined problem of derivation and of formation [13]. They have achieved their initial goal of developing a program that could produce reasonable problem-solving trees containing a high density of complete and satisfactory syntheses [12]. They are currently extending the scope of the program by including features such as stereochemistry. More realistic heuristic rules for tree-pruning are being added.

Although the details of programming techniques are different, this program is essentially similar to LHASA and the comments made about LHASA in the last paragraph of Section A are applicable to it.

C. Hendrickson's Method

A less empirical approach to synthesis analysis has been taken by Hendrickson [14]. He recognizes the need for a system of characterization of structures and reactions that is specifically designed for use in synthesis. Obviously such a system should derive from the fundamentals of structure, and should be general enough to allow the inclusion of new reactions without collapse of the system. It should convey the information a chemist looks for first when he considers a molecule that is to be synthesized, *i.e.* the carbon skeleton and functionality. Finally, the system should be so designed that it takes into account, at least implicitly, *all possible* synthetic routes to a given compound.

Since any organic structure is a collection of carbon sites, and any reaction is a change in the characteristics, *i.e.* functionality and skeletal position, of one or more of these sites, it is convenient to define a carbon site by its four attachments: $R = \sigma$ bond to carbon, $\Pi = \Pi$-bond to carbon, $H =$ bond to H or other less electronegative atoms, $Z =$ bond to more electronegative atom (*e.g.* bonds to heteroatoms). In addition, Hendrickson defines symbols denoting the numbers of each kind of attachment: $\sigma =$ number of σ bonds to carbon, $\Pi =$ number of Π-bonds to carbon, $h =$ number of bonds to H or other less electronegative atoms, $z =$ number of bonds to heteroatoms. Since a carbon-carbon Π-bond may be considered both as a functional group and a part of the skeleton, it is convenient to define the functionality at a carbon site as $f = \Pi + z$, *i.e.* the number of functional bonds. Obviously,

$$\sigma + \Pi + h + z = f + \sigma + h = 4 \, .$$

Finally, one may define the character of a site $c = 10\,\sigma + f$, a two-digit number in which the first digit shows the skeletal nature and the second digit shows the level of functionality of the carbon site.

Given this definition of character, a simple notation for representing molecular structure can be developed. This consists of a list of site numbers (arbitrary labels given to the various carbon atoms) each with its character as a superscript. Each site number is linked by a dash to the adjacent numbers if they are bonded, or separated by a slash if they are not. Sites other than sequential ones which are attached to a given site are then listed in parentheses after it. This notation is exemplified below for cyclopropanone and camphor.

A. J. Thakkar

$$O$$
$$\|$$
$$C2$$
$$\overset{\wedge}{_{1}C-C3}$$

$1^{20}(3) - 2^{22} - 3^{20}(1)$

Cyclopropanone

$1^{30}(6,7) - 2^{20} - 3^{20} - 4^{40}(7,10) -$
$5^{22} - 6^{20}(1)/7^{40}(1,4,9) - 8^{10}/$
$9^{10}(7)/10^{10}(4)$

Camphor

Given the definitions of the four kinds of attachment to a carbon site, any reaction at that site may be described as the replacement of one kind of attachment by another. Thus a reaction at a single carbon site may be symbolized by a pair of the capital letters (Π, R, H, Z), the first indicating the attachment formed in the reaction and the second the attachment removed. For example, HZ is a replacement of a heteroatom by hydrogen, i.e. a reduction of halide, ketone etc. It is trivially easy to extend this notation to reactions that occur at more than one carbon site. For example, an aldol-type reaction is described as RH.RZ, and a Michael addition as RH.RΠ.HΠ in an obvious notation.

The generality and simplicity of this notation make it an extremely versatile and useful tool. As shown by Hendrickson [14], slight extensions of the notation coupled with the use of elementary graph theory give the system the capability of providing a chemist with a quick overview of all possible reaction *types* which can give rise to a particular structural feature. The character triangle developed by Hendrickson [14] is an example of the utility of the notation. The character triangle can be used to determine whether a proposed synthetic scheme contains any redundancies from a structural point of view. It is also possible to use it to enumerate all the routes of *n* steps to a given site in order to check that all possible *types* of synthetic routes have been considered in a particular case. A major advantage of the system is its simplicity which it makes it adaptable for hand use as well as for computer use.

It would seem that Hendrickson's system of notation is a valuable addition to the impoverished repertoire of tools available to assist a chemist in the planning of syntheses. Hendrickson has already applied it to the synthesis design of substituted aromatics [15].

Another possible use of the system is in the classification of all known chemical reactions into categories relevant to synthesis. Such a classification would show up gaps in the armory of synthetically-useful reactions and thus suggest presently non-existent conversions which might be invented. This

14

was something that was being done [16] before Hendrickson's system was proposed, and hence the latter should serve as an aid to the former task.

Other uses of the system, with its convenient linear notation, may lie in the field of chemical documentation [17] — a field that urgently needs attention in view of the current exponential rate of increase of the chemical literature.

Finally, an extremely important use for Hendrickson's system is a pedagogical one [18]. The system could make a beginning student's path through the maze of synthesis much less tortuous, and is worth implementing at least on an experimental basis.

D. CNPE: The "Non-Empirical" Approach

Another scheme for systematic synthesis analysis has been proposed by Ugi and Gillespie [19]. The system is based on the recognition that all chemical reactions correspond to interconversions of isomeric ensembles of molecules (IEM) within a family of isomeric ensembles of molecules (FIEM) [20]. They have shown [19] that distinguishable IEM of a FIEM can be represented by a family of be matrices (bond and electron matrices) $F = (M_0, M_1, \ldots M_f)$. The be matrix M_i of an ensemble of molecules, EM_i, consisting of a set A containing n atoms, $A = (A_1, \ldots, A_n)$ is an $n \times n$ matrix as shown below:

$$M_i = \begin{bmatrix} e_1 & i_{12} & i_{13} & \cdots & i_{1n} \\ b_{21} & e_2 & i_{23} & \cdots & i_{2n} \\ b_{31} & b_{32} & e_3 & \cdots & i_{3n} \\ \cdot & \cdot & \cdot & & \cdot \\ \cdot & \cdot & \cdot & & \cdot \\ \cdot & \cdot & \cdot & & \cdot \\ b_{n1} & b_{n2} & b_{n3} & \cdots & e_n \end{bmatrix}$$

where the entries $b_{lm}(l > m)$ are the formal bond orders of the bonds between pairs of atoms A_l and A_m, the diagonal entries $e_k (\equiv e_{ll})$ are numerically equivalent to the number of free valence electrons belonging to atom A_k in EM_i, and the entries $i_{lm} (m > l)$ can represent any property (e.g. thermodynamic, steric etc.) that exists between pairs of atoms A_l and A_m. For mathematical simplicity, the i_{lm} may be set equal to the b_{ml} to obtain a symmetric matrix.

Thus the generation of a synthetic tree reduces to the problem of finding be matrices M_ψ which are related to M_i by the following equation:

$$M_\psi = R_p \cdot M_i$$

This equation ensures, among other things, that the conservation of matter principle is obeyed. Obviously, a computer is used to find the R_p. This scheme has been called CNPE (Complete Neglect of Prior Experince).

It is obvious that this is an extremely general method which is based upon much sounder principles than the methods of Corey and Wipke and, to a lesser extent, those of Hendrickson. However, there are two immediate difficulties encountered in the application of this method. One is that if no criteria are included by which the majority of the R_p are eliminated, then the output from the computer is tremendous in volume, and the majority of routes found are inferior ones. Therefore empirical selection rules have to be included, and these are usually based on information contained in the *i* portion of the **bei** matrices. The other drawback inherent in the use of the principle of conservation of matter is that all side-products and unconsumed starting materials must be specified along with the target molecule. Anyone who is familiar with organic chemistry will realize that this is usually an extremely difficult, if not impossible, task. This shortcoming is patched up by assuming that there are no sideproducts, and then, if this EM_i does not lead to a satisfactory synthetic route, sideproducts are added to the EM_i as needed and the entire program rerun. This procedure is continued until a satisfactory solution is found.

Thus, it seems that in spite of the fact that Ugi's method has the tremendous advantage of not relying upon known chemical reactions, it is not a practical one to use until satisfactory ways of overcoming the difficulties outlined above are found.

However, Ugi's concept of chemical reactions being interconversions of IEM and FIEM is a general one which is likely to find many uses including the interpretation of mass spectral patterns. This work on synthesis analysis is only a small part of Ugi's general effort to bring about greater use of the logical structures of mathematics in chemistry [21-23]. The broader purpose of Ugi's work is indeed commendable since it represents the first steps toward a systematization of the procedures for storing and handling the vast quantity of chemical information that is currently available.

V. Conclusions

Thus, four recent approaches to systematic synthesis analysis have been reviewed and their merits and drawbacks outlined. It seems certain that the use of computer programs to assist synthesis analysis will become routine within the next decade. If anything stands out from the entire discussion, it is that the introduction of computerized methods into the once sacrosanct field of classical synthesis analysis has made the long-predicted computer age of chemistry a reality.

Appendix I: An Algorithm for Efficient Computer Storage of Connectivity Tables

As an illustratration of the types of techniques that may be used to make maximum use of core storage, an algorithm for efficient storage of connectivity tables is presented.

Consider a connectivity table of the type shown in Fig. 2. It is clear that considerable economy of core space can be achieved by storing an entire atom (or bond) entry in a single computer word, with each of the digits serving as an entry in a column of the table.

However, higher-level programming languages such as Fortran IV do not permit direct access to the individual digits of a given word. This problem is easily solved by utilizing the fact that in integer division the remainder is truncated (in Fortran IV). For example, to find the ith digit, D, in a n-digit word, X, we have the following formula

$$D = Y - [Y/10] \cdot 10 \qquad (1)$$

$$\text{where } Y = X/(10)^{n-i} \qquad (2)$$

Obviously Eq. (1) holds only because the remainder is discarded in integer division performed by the computer.

Acknowledgements. I would like to thank Professor G. L. Hofacker for the hospitality extended to me during my stay at the Lehrstuhl für Theoretische Chemie. I would also like to thank Professors Vedene H. Smith, Jr., Saul Wolfe and Ivar Ugi for their encouragement and their helpful comments.

References and Footnotes

[1] Woodward, R. B.: Synthesis. In: Perspectives in organic chemistry, p. 155. New York, N. Y.: Interscience 1956.

[2] Corey, E. J.: Pure Appl. Chem. *14*, 19 (1967).

[3] Corey, E. J., Wipke, W. T.: Science *166*, 178 (1969).

[4] Dr. Cattell has been Research Professor in Psychology and Director of the Laboratory of Personality Assessment at the University of Illinois for the last 25 years. The quotation is from his book: The scientific analysis of personality, p. 102. Penguin: 1967.

[5] For a review see: Corey, E. J.: Quart. Rev. (London) *25*, 455 (1971).

[6] Corey, E. J., Wipke, W. T., Cramer, R. D., Howe, W. J.: J. Am. Chem. Soc. *94*, 421 (1972).

[7] Corey, E. J., Wipke, W. T., Cramer, R. D., Howe, W. J.: J. Am. Chem. Soc. *94*, 431 (1972).

[8] Corey, E. J., Cramer, R. D., Howe, W. J.: J. Am. Chem. Soc. *94*, 440 (1972).

[9] Corey, E. J., Petersson, G. A.: J. Am. Chem. Soc. *94*, 460 (1972).

[10] Woodward, R. B., von E. Doering, W.: J. Am. Chem. Soc. *67*, 860 (1945).

[11] Woodward, R. B., Bader, F. E., Bickel, H., Frey, A. J., Kierstead, R. W.: Tetrahedron 2, 1 (1958).

[12] Gelernter, H., Sridharan, N. S., Hart, A. J., Fowler, F. W., Shue, H. J.: An application of artificial intelligence to the problem of organic synthesis discovery; Technical Report, Departments of Computer Science and Chemistry, State University of New York at Stony Brook, August 1971.

A. J. Thakkar

13) Amarel, S.: Problem solving and decision making by computer: An overview. In: Cognition and artificial intelligence: A multiple view (ed. Garvin). Aldine Press 1969.
14) Hendrickson, J. B.: J. Am. Chem. Soc. *93*, 6847 (1971).
15) Hendrickson, J. B.: J. Am. Chem. Soc. *93*, 6854 (1971).
16) The discovery of new methods for performing specific structural operations requires first a *realization* of certain unfilled needs in the field of synthesis. Quoted from Ref.[2], p. 31. See Ref.[2] for a review of such work.
17) See for example: Morgan, H. L.: J. Chem. Doc. *5*, 107 (1965). For a review see Ref.[3].
18) See the text: Ireland, R. E.: Organic synthesis for a representative example of currents methods of teaching synthetic planning. Prentice-Hall, N. Y. 1969.
19) Ugi, I., Gillespie, P.: Angew. Chem. Intern. Ed. Engl. *10*, 915 (1971).
20) Ugi, I., Gillespie, P.: Angew. Chem. Intern. Ed. Engl. *10*, 914 (1971).
21) Gillespie, P., Hoffmann, P., Klusacek, H., Marquarding, D., Pfohl, S., Ramirez, F. Tsolis, E. A., Ugi, I.: Angew. Chem. Intern. Ed. Engl. *10*, 687 (1971).
22) Ugi, I., Marquarding, D., Klusacek, H., Gokel, G., Gillespie, P.: Angew. Chem. Intern. Ed. Engl. *9*, 703 (1970).
23) Dugundji, J., Gillespie, P., Ugi, I., Marquarding, D.: The metric nature of chemistry. In: The chemical applications of graph theory (ed. A. Balaban). Academic Press to be published

Received October 6, 1972

An Algebraic Model of Constitutional Chemistry as a Basis for Chemical Computer Programs

Prof. Dr. James Dugundji

Department of Mathematics, University of Southern California, Los Angeles, California, USA

Prof. Dr. Ivar Ugi

Department of Chemistry, University of Southern California, Los Angeles, California, USA and Laboratorium für Organische Chemie der Technischen Universität München

Contents

Summary

The presently available chemical computer programs are discussed, as well as the fact that a mathematical theory of constitutional chemistry is needed as a basis for substantial progress in computer assisted solutions to chemical problems. The representation of EM [1] and their interconversions by be-matrices [1] and reaction matrices leads to the logical structure of the FIEM [1] and is representative for constitutional chemistry. A mathematical theory of the FIEM is stated, which can serve as the foundation of a new type of matrix-fitting computer programs for establishing constitutional relations such as synthetic design, the elucidation of reaction mechanisms, and bio-synthetic pathways, as well as the analysis of mass spectra.

Zusammenfassung

Die gegenwärtig verfügbaren chemischen Computer-Programme werden erörtert, sowie die Tatsache, daß wesentlicher Fortschritt bei der Lösung chemischer Probleme mittels Computern eine mathematische Theorie der konstitutionsbezogenen Chemie erfordert. Die Darstellung von EM [1] und deren wechselseitigen Umwandlungen durch be-Matrizen [1] und Reaktions-Matrizen führt zur Einsicht, daß die logische Struktur der FIEM [1] repräsentativ ist für die konstitutionsbezogene Chemie. Eine mathematische Theorie der FIEM wird angegeben. Diese kann als Grundlage einer neuen Art von Computer-Programmen zur Ermittlung konstitutioneller Beziehungen dienen, wie z. B. Synthese-Planung, Aufklärung von Reaktions-Mechanismen und Stoffwechselvorgängen, sowie Auswertung von Massen-Spektren.

21

I. The Basic Types of Scientific Computer Programs

The use of computers has dramatically expedited progress in many fields of science. There are various scientific problems whose effective solution is only possible with the aid of computers.

Two types of computer programs serve the purpose of science: the *logical structure oriented* and non-empirical programs that are based upon mathematical structures and formalisms, and the *information oriented* and empirical programs that rely primarily on input data and use algorithms[a] which are essentially derived from empirical information to process the input data.

Among the chemical applications of computers only very few are predominantly logic oriented, such as the use of computers in x-ray crystallography [2] and the computation of quantum mechanical data, sometimes with an accuracy that, in many cases, suffices to predict some of the physical and chemical behavior of molecules.[3] The advances in chemical documentation by computer storage and retrieval of suitably coded chemical information rely on information oriented programs.[4] The recent spectacular successes in computer-assisted systematic evaluation of vast amounts of chemical and physicochemical experimental results, is accomplished by computer programs which use mathematical structures and formalisms to some extent, but are still mainly information oriented.[5, 6]

A further classification of computer programs is possible on the basis of the role which the operator plays:

There are the *batch programs* that are able to generate the final results from the input without any intermediate participation of the operator, and there are the *dialog programs* where the operator must make some of the intermediate decisions on the basis of intermediate data output.[7, 8]

As a rule, the dialog programs are more time consuming and less effective than the batch programs. Often the development of a batch program goes through the stage of a dialog program.

The use of computers in chemistry appears to be limited to "peripheral" and auxiliary applications, because hitherto, chemistry itself seemed not to be translatable into mathematics to an extent which is required for the effective use of computers in the direct solution of complex purely chemical problems by logic oriented programs: the set of data which would be required for a widely useful information oriented computer program for the solution of chemical problems is just too big and heterogeneous to be manageable, even for the most advanced of the available computers. Furthermore, for many chemical problems, empirical information in combination with simple mapping procedures does not suffice for generating non-trivial solutions by a dialog program.

---

[a] An algorithm is a list of instructions specifying a sequence of operations which will give the answer to a problem of a given type.

Chemistry is still a science which also contains some elements of an art, and a present-day chemist who is planning a synthesis, or is elucidating a reaction mechanism, needs not only logical abilities but also experience and intuition.

In general, satisfactory approximations to complete sets of solutions of chemical problems (such as synthesis) cannot be obtained with the aid of artificial intelligence, unless a unified mathematical theory of chemistry becomes available which can be used in a simple and effective manner as a basis of computer programs. Any purely empirical, heuristically guided[8)b)] information oriented trial-and-error search program which does not rely on a mathematical theory of the structure of chemistry, can only be applied to narrow areas of interest, generating information which is closely related to the input data, *i. e.*, mostly some small subset in a space of possible solutions to a problem. All of the existing computer programs for synthetic design are of the information oriented type[c)], and are confined to certain classes of synthetic targets and types of known reactions. This is equally true for the *Harward dialog type program* [9)] for mainly hydrocarbons and CHO compounds, *e. g.*, terpenes, the Princeton program [*], the *Johns Hopkins aromatic compounds program* [10)], the *program for Diels-Alder and Cope reactions* and analogs at Rice [11)], as well as the *batch type programs*, such as the Leverkusen program for peptides and proteins (see below) [12)], and also one of the most advanced of all, the quite versatile *Stony Brook PL/1*

---

b) By some "rules of thumb", or trick, that serve to aid discovery.

c) The Harvard (LHASA) program with emphasis on the use of graphic devices has been termed "logic centered". However, by the above criteria, it is strictly information oriented. The term "logic centered" is used just to indicate that a systematic search for precursors by some algorithm is involved. This program is based upon a pattern recognition routine that is applied to the synthetic target molecule or its subtarget. The latter is dissected to yield precursors by mapping operations which are selected from an input of the "inverses" of a set of known synthetic reactions (transforms) according to the target pattern. These mapping instructions may be extremely complicated parts of the program; the Diels-Alder transform of the Harvard-Princeton program required 6 months of programming time alone. For example, perception of the pattern $-CHOH-CH_2-CO-$ leads to application of the transform of the aldol condensation which belongs to the "list of reactions", and generates a precursor with the pattern $-CHO + CH_3-CO-$. The precursors are displayed graphically on a CRT/ and are inspected by the operator (dialog program), or evaluated, according to listed yield estimates of reactions that would produce the target from the precursor. The precursor favored by the evaluation is then, again, used as a target, etc., etc., until starting materials are reached which are acceptable to the operator.

* The *Princeton FORTRAN program* with graphic emphasis was developed by a group around W. T. Wipke, an initiator of the Harvard program [9)], as a more advanced version of the latter. This program includes even some stereochemical capability, based upon classical potentials.

program [13], which includes a sophisticated tree search procedure using a heuristic evaluation of the nodes and refers to a *Wiswesser* coded supplemented *Aldrich Catalog* as a "starting material library".

II. The Need for a Unified Structural Theory of Constitutional Chemistry

The attempt to find any approximation to the solution of some problem for which the complete and perfect solution cannot be obtained, requires nevertheless some ideas about the desirable perfect solution. The ideal computer program for synthetic design would be a logic oriented non-empirical batch program and would produce for a given target molecule Z a small number of synthetic pathways which, on the basis of predetermined criteria, are the best methods to synthesize Z among all possible methods for doing so.

Without a logical structure of constitutional chemistry, one might conjecture that for any synthetic target Z there is an infinite tree of synthetic pathways [14] whose nodes can never be fully considered by a synthetic design program. This is true only if one distinguishes between all syntheses that differ in any respect, including the reaction conditions of each step involved.

A chemical synthesis of a target molecule could be described in terms of the empirical formulas of the starting materials and some representative intermediates which are characteristic of the conversion of the starting materials into the synthetic target. Such a description would, however, be useless for most synthetic purposes, because it would be representative of a too-large equivalence class of syntheses. Perkins' attempt to synthetize quinine from aniline, which led to the discovery of the triphenyl methane dyes, is a well documented example [15] for the need of defining and representing equivalence classes of syntheses in a more detailed manner than by some characteristic empirical formulas.

The number of conceivable synthetic pathways is reduced substantially if we confine the term *synthesis* to those reasonable sequences of chemical reactions which lead from some readily available starting materials to the desired target compounds without obvious waste of material and labor, and without involving nonsense sequences of operations, *i. e.*, synthetic pathways that contain unnecessary "loops".[d] For instance, we should not consider a synthesis of NaCl from strychnine hydrochloride and triphenylmethyl sodium [16] a reasonable synthetic of NaCl.

---

[d] The recognition of (non-cyclic) "loops" requires, in fact, some knowledge of structure of constitutional chemistry.

There are a limited number of immediate precursors for any synthetic target which itself may be a precursor of another target, and there are only a rather small number of chemical compounds which are habitually used as starting materials of any synthesis. The catalogs of the major suppliers of chemicals contain a rather complete set of the synthetic starting materials that are presently being used. Anybody would agree immediately that there is only a finite number of reasonable syntheses for sodium chloride, or also methanol, from listed starting materials. There are also only a finite number of constitutional synthetic pathways (see below) for vitamin B_{12}, the most complex known "small molecule"; if this were not true, there would be a certain limit of molecular complexity, below which the number of conceivable reasonable syntheses is well-defined, and above which, there is an infinite number of synthetic alternatives.

It follows from the theory of constitutional chemistry which is presented in this paper that for any target molecule there exists an upper bound for the number of *constitutional synthetic pathways*. Since any of the latter contain a well-defined number of stereochemical synthetic pathways, the number of stereochemical pathways for a given synthetic target also has an upper bound.[17] In view of the practical limits on the amount of time and storage available, for some complex synthetic targets, these numbers may be too large for exhaustive blind search methods for finding paths to a goal node, because the search will explore too many nodes; however, there are state space search methods [8], such as dynamic programming [18] which can be used to find, according to some chosen criteria, the best multistep synthetic pathways within sets of conceivable syntheses without the individual evaluation of each pathway.[12]

Already in the mid-sixties, the model experiments for peptide syntheses by stereoselective four component condensations demonstrated the potential usefulness of artificial intelligence [5] for the solution of synthetic problems. The optimum reaction conditions for the stereoselective synthesis of a given four component condensation product can be found only with some rather detailed knowledge of the very complex reaction mechanism of four component condensations. The latter could be elucidated only with the aid of a computer.[12,19]

This led further to preliminary computer assisted planning study of the synthesis of a variety of proteins, such as the A chain of insulin, by a combination of conventional peptide synthetic methods and stereoselective four component condensations which revealed that, for the latter, there are approximately 139,000,000 syntheses of the above-mentioned type, differing widely with respect to yield expectations and number of required steps.[12] Although it would be beyond the capabilities of the presently available advanced computers to generate and compare all of these conceivable pathways according to suitable criteria, it took only a few minutes of computing time with an IBM 7094 computer to generate the ten "best" syntheses for insulin A when the dynamic programming technique and evaluation by inverse yield functions [12] was applied. In the latter case, the conditions for the effective use of a computer for synthetic planning are uniquely favorable, because large numbers of operations are involved which belong to only a few different types of reactions, and for each synthetic target the set of starting materials is easy to determine.

The design of a general synthetic design program which is not confined to certain classes of compounds, and is capable of taking into account unprecedented, but conceivable, chemistry, requires a unified mathematical theory of constitutional chemistry on the basis of a few general principles.

Once there is a theory which permits the manufacture of representatives for all precursors and reaction steps of all conceivable synthetic pathways for a given target, then it will be possible to develop a computer program which, considering the set of all pathways which do not violate the input principles, and which, without manufacturing and evaluating these pathways individually, produces with justifiable amounts of computer time representations for a chosen number of the best syntheses for a given target molecule.

III. The Logical Structure of Chemistry

The physical nature of the molecules and their subunits suggests that, in principle, the logical structure of chemistry could be derived from quantum mechanics. This would, however, require methods by which the potential energy surface of any system containing the same set A of n atoms could be analyzed.

An attempt in this direction has, in fact, been made by Preuss [20], who recognized the importance of the above problem and developed methods for the treatment of the *"incomplete associations of atoms"* which contain the FIEM (see Section A.) as subsets. The complexity of the Preuss approach limits its application to systems which contain much fewer nuclei and electrons than those involved in complicated chemical problems, such as synthetic design. For the latter, a different type of theory is needed, a theory that affords insights into the relations between complex chemical systems, without the sometimes formidable effort of solving the pertinent quantum mechanical problems. It is, however, conceivable that the potential energy surfaces of polyatomic systems will be the basis of future computer programs for the solution of chemical problems.

The logical structure of chemistry is separable into the logical structure of *stereochemistry* [17,21-26] and the logical structure of *constitutional chemistry*.[17,27] Since the stereochemical systems are representable in terms of skeletal symmetries and distributions of permutable ligands, group theory is a suitable device for the theoretical treatment of stereochemistry.[21,23-26] The logical structure of static and dynamic stereochemistry has been elucidated by group theoretical methods and is representable by the subgroups, cosets, double cosets, classes and subclasses of the permutation groups S_n.[24-26]

Our present knowledge of constitutional chemistry indicates that there exists some universal principle of order to which the chemical systems and their relations are subject. What concepts do we need, and how do we have to define the chemical systems and their relations in order to be able to spell out this universal structure?

Two generally accepted ideas provide the basis for achieving this: the principle of *conservation of matter* during chemical transformations and the approximation standpoint that the atoms which participate in a molecular structure can be treated in terms of a *core* A_i composed of the nucleus and tightly bound inner electrons with a positive formal core charge $e_{i,0}$ together with a *valence shell of electrons*[3] that are less tightly bound, and are responsible for the *chemical bonds*, such as the ionic bonds, the covalent electron pair bonds, and the multicenter bonds.

It is the objective of the present paper to deal with the elucidation of the structure of constitutional chemistry and its potential usefulness for the solution of chemical problems, in particular, for the design of syntheses.

A. Chemistry of a Fixed Set of Atoms, and the FIEM

Extension of the equivalence relation of *isomerism* to *ensembles of molecules* leads to a unified theory for the relations between chemical systems. Previously, the potential importance of this extended equivalence relation to the interpretation of chemistry and systematic planning of chemical experiments seem to have been overlooked.

There are two types of empirical formulas for EM, the *gross empirical formula* which indicates the total number of atoms that are contained in an EM, and the *detailed empirical formula* which represents the set of empirical formulas of the molecules that belong to the EM.

Let A be a given set of atoms. The empirical formula of A is the gross empirical formula of all *isomeric* EM that can be formed from A. The detailed empirical formula of an EM(A) is a partition $\{\overline{A}_1, \ldots, \overline{A}_s\}$ of A (*i.e.*, the $\overline{A}_i$ are pairwise disjoint and $A = \bigcup_1^s A_i$) such that each A_i is the empirical formula of a molecule. Thus, an EM(A) is any compound or collection of chemical species that can be formed from A using each atom belonging to A exactly once. Since isomeric molecules can differ constitutionally, there can be more than one EM(A) with a given detailed empirical formula. The constitutional formula of an EM(A) is the set of formulas of all the molecules belonging to that EM(A).

An FIEM(A), the *family of isomeric ensembles of molecules* of A, is the set of all EM(A).[17] Thus, an FIEM is described simply by a gross empirical formula; an EM is described by a list of molecular species in terms of constitutional formulas.

The constitutional chemistry of a set of atoms A is given by the FIEM(A). We elaborate this in more abstract terms:

Each finite set A of atoms can be described by an empirical formula, which lists the number, and type, of atoms present in A. We let E be the set of all empirical formulas. Let M be the family of all finite sets of chemical species, so that a member of M is a set of chemical species. To each $\mu \in M$, we associate an element $p(\mu) \in E$, which gives the total number, and type, of atoms that occur in μ. This determines a map $p : M \to E$. For each $A \in E$, the set $p^{-1}(A)$ is called the FIEM(A); an EM(A) is any member of the FIEM(A). Thus, the FIEM(A) is the family of all sets of chemical species having empirical formula A; any particular member of this family is called an EM(A).

The chemical constitution of an EM(A) is determined by the set of the cores of A and the distribution of the valence electrons. In our present model concept the valence electrons are either pairwise shared by pairs of cores and form *covalent bonds*, or occupy valence orbitals of individual atoms as *lone electrons*. Thus the chemical constitution of an EM(A) is given by its covalent bonds and bonded neighbors, and its lone valence electrons. Conventionally, this is represented by the constitutional chemical formulas.

The chemical constitutions of EM, and the constitutional relations between EM, define the fundamental logical structure of chemistry within the FIEM. The chemical processes are the interconversion of EM(A) by redistributions of the valence electrons between the cores; the left and right sides of chemical equations refer to isomeric EM: reactions and sequences of reactions correspond to interconversions of EM belonging to the same FIEM. For example, the starting materials and the products of Reaction 1 have both the gross empirical formula CH_3NO (so belong to the same FIEM); the detailed empirical formula (*i. e.*, the EM) of the starting materials is, however, $CHN + H_2O$.

Reaction 1

$$H-C \equiv N: + H-\overset{..}{\underset{..}{O}}-H = H-\overset{..}{\underset{\underset{:O:\ H}{\|\ \ \ |}}{C}}-N-H$$

$$\textit{1} \qquad\qquad \textit{2a} \qquad\qquad \textit{3}$$

In a chemical equation the equality sign expresses a stoichiometric equivalence relation.

Reaction 2, the further hydrolysis of *3*, corresponds to an interconversion of isomeric EM with the gross empirical formula CH_5NO_2, by redistribution of the valence electrons.

Reaction 2

$$H-\overset{\overset{\displaystyle\|}{O:}\ \ \overset{\displaystyle|}{H}}{C}-\overset{..}{N}-H \ + \ H-\overset{..}{\underset{..}{O}}-H \ = \ H-\overset{\overset{\displaystyle\|}{O:}}{C}-\overset{..}{\underset{..}{O}}:^{\ominus} \ + \ H-\overset{\overset{\displaystyle H}{|}}{\underset{\displaystyle|}{N}}{}^{\oplus}-H$$

$$\quad\quad 3 \quad\quad\quad\quad 2b \quad\quad\quad 4 \quad\quad\quad 5$$

By adding *2b* as a complement to Reaction 1, it enables us to describe both Reactions 1 and 2 in terms of interconversions of isomeric EM, according to Reaction 3.

Reaction 3

$$EM_1 \longrightarrow EM_2 \longrightarrow EM_3$$

with:

$$EM_1 = \{1 + 2a + 2b\}$$
$$EM_2 + \{3 + 2b\}$$
$$EM_3 = \{4 + 5\}$$

A multistep synthetic pathway, just as any sequence of chemical reactions can be represented as a sequence of EM transformations within an FIEM.

The translation of conventionally represented syntheses into synthetic pathways within an FIEM and *vice versa* requires determining the pertinent EM for each stage of each synthetic pathway. The EM which represents a given stage of a synthetis contains the starting materials that have not been consumed at that stage, its characteristic intermediates which have already been formed, but have not undergone yet further reactions, as well as all products and byproducts which already exist and do not participate in further transformations.

This is illustrated by the following two examples:

The "linear synthesis" according to the stepwise strategy of Scheme 1 corresponds to the Pathway 1.

Scheme 1

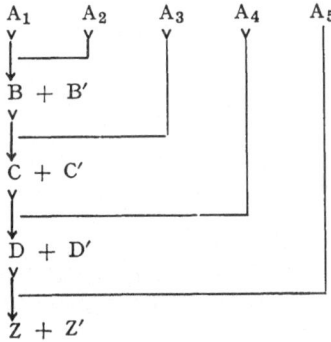

J. Dugundji and I. Ugi

Pathway 1

$$EM_A \longrightarrow EM_B \longrightarrow EM_C \longrightarrow EM_D \longrightarrow EM_Z$$

with: $EM_A = \{A_1 + A_2 + A_3 + A_4 + A_5\}$
$EM_B = \{A_3 + A_4 + A_5 + B' + B\}$
$EM_C = \{A_4 + A_5 + B' + C' + C\}$
$EM_D = \{A_5 + B' + C' + D' + D\}$
$EM_Z = \{B' + C' + D' + Z' + Z\}$

The "branched synthesis" according to the fragment strategy [28] of Scheme 2 corresponds analogously to Pathway 2.

Scheme 2

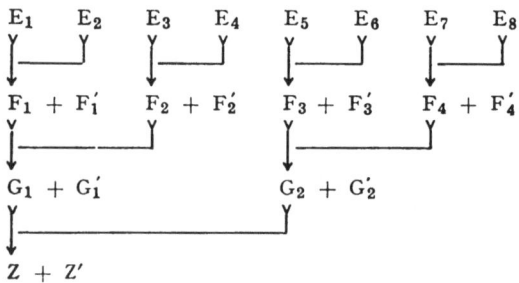

Pathway 2

$$EM_E \longrightarrow EM_F \longrightarrow EM_G \longrightarrow EM_Z$$

with: $EM_E = \{E_1 + E_2 + E_3 + E_4 + E_5 + F_6 + E_7 + E_8\}$
$EM_F = \{F_1' + F_2' + F_3' + F_4' + F_1 + F_2 + F_3 + F_4\}$
$EM_G = \{F_1' + F_2' + F_3' + F_4' + G_1' + G_2' + G_1 + G_2\}$
$EM_Z = \{F_1' + F_2' + F_3' + F_4' + G_1' + G_2' + Z' + Z\}$

We use the term synthetic pathway for a transformation of a set of starting material molecules EM_A via a sequence of isomeric intermediate EM into an EM_Z which contains the synthetic target molecule Z and its by-products.

Pathway 3

$$EM_A \longrightarrow EM_B \longrightarrow EM_C \longrightarrow \ldots \longrightarrow EM_Y \longrightarrow EM_Z$$

We consider any two synthetic pathways as equivalent syntheses of Z if one can be obtained from the other by adding (or deleting) isomeric EM to the EM of each step. Thus, pathway 5, which is obtained by combination of pathways 3 and 4, is as a synthesis of Z equivalent to pathway 3 [3].

Pathway 4

$$EM_A' \longrightarrow EM_B' \longrightarrow \dots \longrightarrow EM_Y' \longrightarrow EM_Z'$$

Pathway 5

$$EM_A \ EM_A' \longrightarrow EM_B \ EM_B' \longrightarrow EM_Y \ EM_Y' \longrightarrow EM_Z \ EM_Z'$$

since 3 is obtained from 5 by neglecting the isomeric EM˙ which do not directly participate in the synthesis of EM_Z.

The translation of a synthetic pathway of EM within an FIEM into a conventional sequence of chemical equations is accomplished as follows:

The formation of the target EM_Z from its immediate precursor EM_Y is written as a chemical equation $EM_Y = EM_Z$. If EM_Y contains a subset $\overset{\circ}{Y}$ of molecules which have no atoms in common with the target molecule Z this subset $\overset{\circ}{Y}$ is eliminated from EM_Y, the left-hand side of the equation, and a balancing subset $\overset{\circ}{Z}$ is cancelled from EM_Z to give a reduced chemical equation, $\overline{EM}_Y = \overline{EM}_Z$.

Subsequently, the chemical equation containing EM_Y and its precursor EM_X is reduced to $\overline{EM}_X = \overline{EM}_Y'$, such that $\overline{EM}_X$ contains only molecules which have at least one atom in common with some molecule $\in \overline{EM}_Y$. This reduction procedure is successively applied to the precursor of EM_X, etc., etc., until an EM_A is reached which contains only acceptable starting materials for the synthesis of Z. By addition of all reduced chemical equations of a synthetic pathway and cancellation of all species which occur on both sides of the resulting equation, one obtains the gross chemical equation $(\widehat{EM}_A = \widehat{EM}_Z)$ of a synthetic pathway for Z. The gross empirical formulas of $\widehat{EM}_A$ and $\widehat{EM}_Z$ correspond to a set of atoms A whose FIEM(A) is the smallest FIEM that contains the given synthetic pathway. This pathway can also be imbedded in some FIEM(A^*), $A^* \supset A$.

For each target molecule Z there is a set of atoms A whose FIEM(A) contains all conceivable synthetic pathways for Z. The synthetic pathways of Z in its universal FIEM(A) of syntheses are representable by a graph, the universal tree of syntheses of Z. A perfect computer program of synthetic design would find the best synthesis of Z which is contained in the universal tree.

Any two synthetic pathways whose EM are stereoisomeric are called constitutionally equivalent, and the equivalence class of constitutionally equivalent synthetic pathways is termed a constitutional synthetic pathway. Any pathway belonging to a given constitutional pathway is representative of the latter. These concepts will be made more precise after we discuss *be*- and reaction-matrices (cf. Section F).

Reaction mechanisms, sequences of mass spectroscopic processes and biochemical pathways can, in a similar way, be imbedded in the pertinent FIEM.

31

B. be-Matrices

The covalent bonds of a chemical constitution of an EM which contains a set of n atoms can be represented by an $n \times n$ ACM. The ACM were introduced in 1963 by Spialter [29] for documentation purposes. In an atom connectivity matrix B, each off-diagonal entry b_{ij} $(i \neq j)$ indicates the formal bond order between an indexed pair of atoms, A_i and A_j. The ACM are symmetric matrices because the chemical bond is a symmetric relation.

The ACM are not quite adequate for the representation of chemical reactions, since the latter involve not only redistributions of bonds but also unshared valence electrons that do not participate in bonds between atom pairs of the individual molecules. The adequate representation of interconvertible chemical systems requires a device with *double bookkeeping capability*, i. e., matrices which account for both, *bonds and electrons*. The be-matrices are obtained from the ACM by inserting diagonal entries b_{ii} which indicate the number of *free*, unshared valence electrons of the atom A_k. Clearly, each be-matrix B is symmetric, so that the entries on the diagonal and in the upper triangle determine the matrix B uniquely.

The set of all the EM of a FIEM can be represented by a family $F = B_0$, ..., B_f of be-matrices. Each be-matrix contains all the constitutional information of the EM, i. e., all information concerning the bonds and certain aspects of valence electron distributions in that EM, which conventional chemical formulas would contain. The be-matrices can be considered representations of density matrices (see also Section III.C).

The representation of EM and their transformations by be-matrices is illustrated by the be-matrices B of $1 + 2$ and E of $4 + 5$ and the reaction matrix (see Section E).

$$H-C\equiv N: \ + \ H-\overset{..}{\underset{..}{O}}-H \ + \ H-\overset{..}{\underset{..}{O}}-H \ \longrightarrow \ H-C-\overset{..}{O}:^{\ominus} \ + \ H-\overset{H}{\underset{H}{\overset{|}{N}}}-H$$

$$\qquad\qquad\qquad\qquad\qquad\qquad\qquad\qquad \underset{\displaystyle :O:}{\|}$$

| | 1 | | 2a | | 2b | | 4 | | 5 |

$$B = \begin{bmatrix} 0 & 1 & 0 & 0 & 0 & 0 & 0 & 0 & 0 \\ 1 & 0 & 3 & 0 & 0 & 0 & 0 & 0 & 0 \\ 0 & 3 & 2 & 0 & 0 & 0 & 0 & 0 & 0 \\ 0 & 0 & 0 & 0 & 1 & 0 & 0 & 0 & 0 \\ 0 & 0 & 0 & 1 & 4 & 1 & 0 & 0 & 0 \\ 0 & 0 & 0 & 0 & 1 & 0 & 0 & 0 & 0 \\ 0 & 0 & 0 & 0 & 0 & 0 & 0 & 1 & 0 \\ 0 & 0 & 0 & 0 & 0 & 0 & 1 & 4 & 1 \\ 0 & 0 & 0 & 0 & 0 & 0 & 0 & 1 & 0 \end{bmatrix} \begin{matrix} H \\ C \\ N \\ H \\ O \\ H \\ H \\ O \\ H \end{matrix} \qquad E = \begin{bmatrix} 0 & 1 & 0 & 0 & 0 & 0 & 0 & 0 & 0 \\ 1 & 0 & 0 & 0 & 2 & 0 & 0 & 1 & 0 \\ 0 & 0 & 0 & 1 & 0 & 1 & 1 & 0 & 1 \\ 0 & 0 & 1 & 0 & 0 & 0 & 0 & 0 & 0 \\ 0 & 2 & 0 & 0 & 4 & 0 & 0 & 0 & 0 \\ 0 & 0 & 1 & 0 & 0 & 0 & 0 & 0 & 0 \\ 0 & 0 & 1 & 0 & 0 & 0 & 0 & 0 & 0 \\ 0 & 1 & 0 & 0 & 0 & 0 & 0 & 6 & 0 \\ 0 & 0 & 1 & 0 & 0 & 0 & 0 & 0 & 0 \end{bmatrix} \begin{matrix} H \\ C \\ N \\ H \\ O \\ H \\ H \\ O \\ H \end{matrix}$$

$$\quad\;\; H \ C \ N \ H \ O \ H \ H \ O \ H \qquad\qquad\qquad H \ C \ N \ H \ O \ H \ H \ O \ H$$

The sum b_k over the kth row or column of a be-matrix B is the number of valence electrons that belong directly to the atom A_k in that EM. The number b_k lies within an interval $\langle b_{k,min}, b_{k,max} \rangle$ which is characteristic of the chemical element E_k to which the atom A_k belongs, and the chemistry under consideration, so that

$$b_{k,min} \leq b_k = \sum_{i=1}^{n} b_{ik} = \sum_{j=1}^{n} b_{kj} \leq b_{k,max}$$

The formal electrical charge of A_k in the given EM is $b_k^\oplus = b_k - b_{kk}^0$, where b_{kk}^0 is the number of valence electrons of the free atom A_k. Note that b_{kk}^0 is equal to the kth diagonal entry of the be-matrix of the set of free atoms $A_0(n)$.

The total number $\hat{b}_k = 2b_k - b_{kk}$ of valence electrons that are associated with A_k lies within an interval $\langle b_{k,min}, b_{k,max} \rangle$, which is characteristic of A_k and the chemistry under consideration, so that

$$\hat{b}_{k,min} \leq \hat{b}_k \leq \hat{b}_{k,max}$$

The sum T over all entries is equal to the total number of valence electrons of the EM, and corresponds to the number T_0 of valence electrons in $A(n)$,

$$T = \sum_{k=1}^{n} b_k = \sum_{k=1}^{n} b_{kk}^0$$

Two be-matrices represent the same EM if they are interconvertible by permutations atom indices, i. e. rows/columns.

An EM containing m distinct molecules can be represented by any one of $m!$ be-matrices in block form, where each block represents a distinct molecule of the EM.

The strongly contributing valence bond structures of resonance systems with delocalized π-electrons are represented by a class of be-matrices which differ only in bond orders and diagonal entries but not in connectivities, i. e., if one replaces all non-zero b_{ij} in these be-matrices by "ones" and the diagonal entries by "zero", they all yield the same adjacency matrix.

C. Multicenter, Integral, Closed Shell, and Restricted Chemistry

There is a variety of chemical species whose constitution cannot be adequately described in terms of electron pair bonds between pairs of cores. Examples are the boron hydrides, the "transition" metal complexes of hydrocarbons and their anions, such as π-allyl nickel or ferrocene, or the hypothetical intermediate of the limiting $S_N 2$ mechanism [26]. In these cases the molecular structure can be adequately represented by the model concepts of *multicenter bonds* and fractional bond orders.

EM with fractional bond orders can be represented by be-matrices having rational entries. The off-diagonal entries are

$$b_{ij} = x_{ij}/y_{ij}; \quad \text{here} \quad x_{ij} = 1, 2, 3, \ldots$$

(in most cases $x_{ij} = 2$) is the number of valence electrons that belong to the multicenter bond in which A_i and A_j participate, and $y_{ij} = 2, 3, \ldots$ the number of cores that share the electrons.

Even if only inadequate information on the chemical constitution of a species is available, the double bookkeeping properties of be-matrices avoid serious errors in the algebra of the rational be-matrices and their reaction matrices.

For example, the chemical constitution of diborane [30] is rather described by 6 than by 7,

however, both can be used for the rational be-matrices B, but 7 with $\frac{1}{2}$ bonds is more convenient, than 6 with $1/3$ and $2/3$ bonds and formal $1/3$ electrical charges on two of the hydrogens and the borons.

The rational $n \times n$ be-matrices B can be converted into equivalent $n \times n$ matrices with integral entries by multiplication with the smallest common denominator of the fractional bond orders. By this the mathematical properties of the integral be-matrices and of their reaction matrices can also be used for chemical systems with fractional bond orders.

The chemistry of those molecular systems with electron pair covalent bonds is called *integral chemistry*. All entries of the be-matrices of integral chemistry are integers.

With few exceptions, in the stable organic compounds, all atoms have *closed shells*; the valence shells of all atoms are closed, *i.e.*, singlet states with equal numbers of electrons having α and β spin, and can be approximated by an orbital description in which each orbital is either doubly occupied or empty. The be-matrices of closed shell EM have even diagonal entries $(b_{kk} = 0, 2, 4, 6)$.

We shall use the term *restricted chemistry* for the chemistry of closed shell molecules which have the following further properties:

(a) None of the atoms carries net electrical charge, *i.e.*, the number b_k of valence electrons that belongs directly to each of the individual atoms A_k is equal to $b_{k,0}$ the formal core charge number. That is, it corresponds to the group index of the given element in the Periodic Table of Elements.

(b) The molecules follow the *Lewis-Langmuir Octett Rule* [31], *i.e.*, all atoms, except H, have a valence shell of one s and three p orbitals filled with overall eight electrons, corresponding to a Ne or Ar configuration; H carries two valence electrons, in a He configuration.

The majority of stable organic molecules consist of the elements, H C, N, O, F, Si, P, S and Cl, and belong to restricted chemistry [32].
The EM of restricted chemistry are represented by restricted be-matrices with $b_k = b_{k,0}$, and $b_k = 8$, except $b_H = 2$.

We have:
Multicenter Chemistry $\supset$ Integral Chemistry $\supset$ Closed Shell Chemistry $\supset$ Restricted Chemistry.

D. Chemical Metric

We define the *chemical distance* between two matrices B, E belonging to the same FIEM by

$$D(B,E) = \sum_{i,j} |e_{ij} - b_{ij}|$$

This is a true metric in the set of matrices belonging to the given FIEM; it is simply the number of electrons involved in an interconversion of B to E, and measures how closely B and E are related with regard to that constitutional aspect.

The chemical metric has also a geometric interpretation.[e] Observe first that any $n \times n$ matrix B corresponds to a point of $\mathbf{R}^{n^2}$ (or to a vector, if $\mathbf{R}^{n^2}$ is regarded as a vector space): specifically, the mapping $B \longrightarrow (b_{11}, \ldots, b_{1n}; \ldots; b_{n1}, \ldots, b_{nn})$ is an embedding of the set of all $n \times n$ matrices into $\mathbf{R}^{n^2}$. (Since be-matrices are symmetric, we could in fact embed the matrices belonging to an n-atom FIEM into $\mathbf{R}^{n(n+1)/2}$). Taking the distance between B, E to be the usual Euclidean distance between their counterparts, we find $d(B, E) = \sqrt{\sum_{i,j} (e_{ij} - b_{ij})^2}$. Now, because $d(B, E) \leq D(B, E) \leq n \cdot d(B, E)$, the chemical distance in the FIEM is essentially equivalent to the Euclidean distance in $\mathbf{R}^{n^2}$ between the counterparts.

Thus, the chemical metric on the EM of an FIEM provides not only a formalism for constitutional chemistry, but also allows us to fully use the *properties of Euclidean spaces in expressing the logical structure* of the FIEM.

E. Matrix Representation of Reactions

Let B be the be-matrix for the starting materials, and E the be-matrix for the end products in a chemical reaction. The chemical reaction $B \to E$ is represented by the *reaction matrix* $R = E - B$. Since it is the difference of two symmetric matrices, every reaction matrix is also symmetric.

---

[e] To the present authors' best knowledge, the term *chemical topology* originates with V. Prelog who introduced it for certain model concepts of the stereochemistry of individual molecules (see Ref. [21] and [34], and Ref. cited therein). The concept of chemical topology as a metric topology whose subsets represent the constitutional aspect of chemical systems and their relations, however, seems to be novel.

The reaction matrix of $1+2a+2b \rightarrow 4+5$ is $R=E-B$ (see above).

$$
R = \begin{bmatrix}
0 & 0 & 0 & 0 & 0 & 0 & 0 & 0 & 0 \\
0 & 0 & -3 & 0 & +2 & 0 & 0 & +1 & 0 \\
0 & -3 & -2 & +1 & 0 & +1 & +1 & 0 & +1 \\
0 & 0 & +1 & 0 & -1 & 0 & 0 & 0 & 0 \\
0 & +2 & 0 & -1 & 0 & -1 & 0 & 0 & 0 \\
0 & 0 & +1 & 0 & -1 & 0 & 0 & 0 & 0 \\
0 & 0 & +1 & 0 & 0 & 0 & 0 & -1 & 0 \\
0 & +1 & 0 & 0 & 0 & 0 & -1 & +2 & -1 \\
0 & 0 & +1 & 0 & 0 & 0 & 0 & -1 & 0
\end{bmatrix}
\begin{matrix}
H \\ C \\ N \\ H \\ O \\ H \\ H \\ O \\ H
\end{matrix}
$$

$$
\ \ H\ \ \ C\ \ \ N\ \ \ H\ \ \ O\ \ \ H\ \ \ H\ \ \ O\ \ \ H
$$

A reaction matrix is essentially a "make-break" indicator of bonds between pairs of atoms, endowed with *double bookkeeping capability for bonds and electrons.*

The off-diagonal entries $r_{ij}(i \neq j)$ indicate how many bonds between A_i and A_j are made (+) or broken (−). The diagonal entries r_{ii} indicates how many free electrons A_i gains (+) or loses (−).

Each reaction matrix represents a class of chemical reactions that have in common the same type of electron redistribution, *i.e.*, a reaction matrix is representative of an "electron pushing arrow pattern." [32f] The non-zero reaction matrices and the *be*-matrices of integral chemistry belong to disjoint subsets of $S(n)$, (see Section IV,c), the additive group of all $n \times n$ symmetric matrices with integer entries.

Accordingly, chemical reactions have some general properties which are independent of the individual reacting systems.

F. Fitting Requirements and the Solution of Chemical Problems

An $n \times n$ reaction matrix R *fits* an $n \times n$ *be*-matrix B if the matrix $B+R=E$ is a *be*-matrix.

In particular, E must be a symmetric matrix with non-negative entries. Since the sum of symmetric matrices is symmetric, the requirement that $E = B+R$ be a *be*-matrix reduces to: for each negative entry $-|r_{ij}|$ of R there must be positive entry b_{ij} of B, such that $e_{ij} = b_{ij} + (-|r_{ij}|) \geq 0$. With the knowledge of the structure and the vector basis of the reaction matrices (see Section IV.C) the set $F(B)$ of reaction matrices whose negative entries fit the non-zero entries of a *be*-matrix B is easy to manufacture.

---

[f] $D(B,E)$ is four times the number of "electron pushing arrows" that represents the conversion of B into E.

In order to refer to chemical systems which are conceivably capable of existence at observable concentrations, some further restrictions must be imposed upon E (see Section B).

The be-matrices of all EM(A) of an FIEM(A) can be manufactured from the be-matrix B of any one of the EM(A) and the set $F(B)$ of all reaction matrices R which fit B.[g] The be-matrices of all EM(A) with T valence electrons correspond to points in R^{n^2} whose chemical distance from the origin is $= T$.

For all reaction matrices $R \in F(B)$ which fit B we have

$$\sum_{ij} |r_{ij}| \leq 2T.$$

Let B and E be equivalent $n \times n$ be-matrices, and let $F(B)$ resp. $F(E)$ be the set of all reactions fitting B resp. E.

$$\text{Let } \boldsymbol{B} = \{B + R | R \in F(B)\} \text{ and}$$

$$\boldsymbol{E} = \{E + R | R \in F(E).$$

Then, using the chemical metric, $F(B)$ is isometric with $F(E)$ and $\boldsymbol{B}$ is isometric with $\boldsymbol{E}$.

The solution of constitutional chemical problems is achieved by a suitable fitting procedure.

For example, we can generate a tree of reasonable synthetic pathways that has to be taken into consideration for an n atom target

$$\text{EM}_Z(A) = Z \cup C(Z)$$

containing the target molecule Z and a set $C(Z)$ of expected by-products,[h] ($e.g.$, $C(Z) = \{3\ H_2O,\ 2\ NaCl,\ CO_2,\ NH_4Cl,\ CH_3OH,\ CH_3CO_2H,\ C_6H_5CH_3\}$ by applying a set $F(B_Z)$ of fitting closed shell $n \times n$ reaction matrices with

$$\sum_{ij} |r_{ij}| \leq 12 \text{[i]}$$

to the be-matrix B_Z of $\text{EM}_Z(A)$. This leads to a set of precursors $\text{EM}_Y(A)$. The latter are again subjected to an analogous set $F(B_Y)$ of fitting reaction matrices, etc., until EM(A) are obtained which contain only acceptable synthetic starting materials.

---

[g] An EM(A) belonging to integral chemistry and the reaction matrices lead to the integral FIEM(A). An EM(A) of closed shell chemistry and the closed shell reaction matrices produce the subset of the closed shell EM(A). An analogous statement applies to restricted chemistry.

[h] The superset of the by-products of the majority of synthetically used chemical reactions is a rather small EM containing less than 30 different compounds.

[i] The upper bounds for the numbers of reaction matrices of this type whose negative entries fit an $n \times n$ be-matrix are:

$$23,150 \text{ for } n = 10,$$
$$193,700 \text{ for } n = 20, \text{ and}$$
$$3,103,750 \text{ for } n = 50.$$

All of the current chemical computer programs (see Sections I and II) yield subsets of the solutions of problems which are based upon the present theory of constitutional chemistry. These subsets refer to certain classes of compounds and known reactions. The present theory can therefore also be used to improve the effectiveness of any one of the current programs (see Section I).

The question of synthesizing a given molecule Z is meaningless unless one also specifies the allowable starting materials. For example, to ask for a synthesis of Z when Z itself is allowed as a starting material, is really not a question at all.

Let, then $\mathscr{L}$ be a list allowable starting materials. Each set L of materials drawn from $\mathscr{L}$ determines a definite EM(L), represented by a be-matrix $B(L)$. If there is a reaction matrix R such that $B(L) + R$ represents an EM that contains the molecule Z, we shall say that the ordered pair (L,R) is a synthesis of Z from L. Since each reaction matrix (as we shall see) is a unique combination of redox and homoaptic-homolytic reactions performed in a certain order, each pair (L,R) gives a synthetic pathway in which each intermediate transformation is one of four elementary types.[1] We shall let

$$\sigma(Z) \; = \; \{(L,R) \mid L \text{ is drawn from } \mathscr{L}, \text{ and } B(L) + R \text{ represents an EM containing } Z\}$$

so that $\sigma(Z)$ is the set of all possible syntheses of Z using starting materials drawn from $\mathscr{L}$.

We now consider two functions on σ (Z) to the non-negative integers

$$(1) \quad M_1(L,R) \; = \; \text{order } B(L)$$

This measures how many atoms are involved in the FIEM within which the synthesis takes place or, stated another way, the "amount of matter" with which we start (clearly, $M_1(L,R) \geq$ order $B(Z)$ always)

$$(2) \quad M_2(L,R) \; = \; D(R)$$

This gives the chemical length of the reaction vector R, so it represents the number of "electron pushing arrows" in the synthesis (L,R) of Z; thus, it represents the "complexity of chemical operations."

The question of "best" synthesis arises. To specify the meaning of "best", it is possible to adopt one of two viewpoints:

(a) "best" is a synthesis starting with minimal matter or

(b) "best" is a synthesis having minimal complexity.

These may not lead to the same result: for example, it is not inconceivable that one can synthesize a Z using 3 reagents, but 10 reactions, whereas with 4 reagents only 5 reactions are needed. Thus the concept of "best" depends on whether (a) of (b) is the primary consideration. Needless to say, in each case (a) and (b) there is no general nor single "best" reaction; there may be many different (L,R) satisfying the chosen definition of "best".

---

[1] Observe that two syntheses (L,R) and (L',R') of Z are equivalent (in the sense used before) if $B(L)$ is a block in $B(L')$ and R is the block in R' acting on $B(L)$.

To codify this, define $M: \sigma(Z) \to \mathbf{Z}^+ \times \mathbf{Z}^+$, by

$$M(L,R) = (M_1(L,R), M_2(L,R))$$

and let

$$m(\mathscr{L}) = \{(L,R) \mid M_1(L,R) = \min\}$$

(the set of syntheses in $m(\mathscr{L})$ are those involving minimal matter). Let

$$p(\mathscr{L}) = \{(L,R) \mid M_2(L,R) = \min\}$$

which is the set of synthesis involving the least possible electron arrow pushing.

Then,

(a) The set of all minimal matter best syntheses of Z starting with $\mathscr{L}$ is

$$\mathrm{MMS}(\mathscr{L},Z) = \{(L,R) \in m(\mathscr{L}) \mid M_2(L,R) = \min\}$$

i.e., the minima of M_2 on $m(\mathscr{L})$ are the least complex of the minimal matter syntheses).

(b) The set of all least complex best syntheses of Z starting with $\mathscr{L}$ is

$$\mathrm{MCS}(\mathscr{L};Z) = \{(L,R) \in p(\mathscr{L}) \mid M_1(L,R) = \min\}$$

i.e., the minima of M_1 on $p(\mathscr{L})$ are those of the least complex synthesis that use the least matter.[k]

IV. The Mathematical Foundations of Constitutional Chemistry

A. Introductory Remarks on Additive Free Abelian Groups

Recall that an (additive) free abelian group A is an abelian group [35] having a basis $\{a_i\}$, *i.e.*, each element of A can be written in one, and only one, way as a linear combination of the a_i with integer coefficients. The numbers of elements in the basis $\{a_i\}$, if finite, is called the rank of A.

We will denote by $\mathbf{Z}^n$ the group whose elements are the lattice points of Euclidean n-space $\mathbf{R}^n$, with the group operation being vector addition; it is a free abelian group of rank n, with $\{(1,0\ldots0), (0,1,0\ldots,0), (0,\ldots,0,1)\}$ as a basis.

We will frequently use the fact that two free abelian groups are isomorphic if, and only if, they have the same rank. Recall that the proof is based on the useful facts

(1) If A has a basis $\{a_i \mid i = 1,\ldots,n\}$ and A' has a basis $\{a'_i \mid i = 1,\ldots,n\}$, then the homomorphism $h: A \to A'$ determined by sending each a_i to a'_i, *i.e.*, $h(\sum \lambda_i a_i) = \sum \lambda_i a_i'$, is an isomorphism, and

(2) If $h: A \to A'$ is an isomorphism, and if $\{a_i \mid i = 1,\ldots,n\}$ is a basis for A, then $\{h(a_i) \mid i = 1,\ldots,n\}$ is a basis for A'.

---

[k] These mathematical criteria of the quality of a synthesis often relate to the overall yield, or number of steps, or cost of labor and materials that are needed to produce a unit amount of the target compound.

It is a standard result that any subgroup C of a free abelian group A is also free abelian. We mention explicitly, however, that a proper subgroup C of A may have the same rank as A (hence C may be isomorphic to A): for example, the group $\mathbf{Z}$ of integers and the proper subgroup $2\mathbf{Z}$ of even integers, both have the same rank, and are isomorphic.

In all that follows, $S(n)$ will denote the additive group of all $n \times n$ symmetric matrices with integer entries. Letting P_{ij} be the $n \times n$ matrix with 1 in the (i,j) position and zero everywhere else, it is well-known, and trivial to verify, that $S(n)$ has

$$\{P_{ij} + P_{ji} \,|\, 1 \leq i < j \leq n\} \cup \{P_{ii} \,|\, i = 1, \ldots, n\}$$

as basis, so it is a free abelian group of rank $\dfrac{n(n-1)}{2} + n = \dfrac{n(n+1)}{2}$

Since $S(n)$ and $\mathbf{Z}^s$, $s = \dfrac{n(n+1)}{2}$, have the same rank, they are isomorphic; an explicit isomorphism $\pi : S(n) \approx \mathbf{Z}^s$ (determined by the obvious correspondence of basis) is given by setting, for each matrix $A = (a_{ij}) \in S(n)$,

$$\pi(A) = (a_{11}, \ldots a_{1n}; a_{22} \ldots, a_{2n}; a_{33}, \ldots a_{3n}, \ldots, a_{nn}).$$

By regarding $\mathbf{Z}^s$ as contained in $\mathbf{R}^s$, the map π gives us a "geometrical" representation of $S(n)$; in particular, $\pi(S(n))$ spans $\mathbf{R}^s$.

B. Properties of be-Matrices

1. Integral be-Matrices

Any be-matrix B is necessarily symmetric because each $i-k$ bond is repeated as a $k-i$ bond in the representation. The diagonal entry b_{kk} gives the number of free valence electrons for the atom k in the particular EM under consideration; and the sum b_k of the entries in the kth row is the total of bonds and free valence electrons for the atom k. Clearly, all entries are non-negative integers. Thus, if we make the

Definition.   Let $B(n)$ denote the set of all $n \times n$ symmetric matrices with non-negative integer entries

then each be-matrix that describes an EM of an FIEM having n atoms will belong to the set $B(n)$.

In this section, we study the algebraic features of $B(n)$. Specifically, we fix an FIEM and study the nature of the be-matrices representing the various EM of that FIEM. Observe that the zero matrix corresponds to the EM that lists all the atomic cores, $i.e.$, atoms of the FIEM without valence electrons.

The set $B(n) \subset S(n)$ is clearly not a subgroup of $S(n)$; in fact, in the geometric interpretation $\pi : S(n) \to \mathbf{R}^s$, we see that $\pi(S(n))$ is contained in a cone in $\mathbf{R}^s$ having the origin as vertex.

Let $G[B(n)]$ denote the smallest subgroup of $S(n)$ containing $B(n)$.

Theorem 1. $G[B(n)] = S(n)$, so that $G[B(n)]$ has rank $\dfrac{n(n+1)}{2}$

Proof. We are to show that, if H is a group containing $B(n)$, then $H = S(n)$. Observe that any $Q \in S(n)$ can be written as the difference $Q = A - B$ of two positive symmetric matrices $(A = \{\max(q_{ij},0)\}, B = \{-\min(q_{ij},0)\})$ so we have $A, B \in B(n)$. Since H must contain A and B, and since H is a group, H contains $A - B = Q$ and this completes the proof.

We know that

$$\{P_{ij} + P_{ji} \mid 1 \le i < j \le n\} \cup \{P_{ii} \mid i = 1,\ldots,n\}$$

is a basis for $S(n) = G[B(n)] \supset B(n)$; in particular, then, this theorem states that, given any FIEM with n atoms, there are precisely $\dfrac{n(n+1)}{2}$ EM such that any EM belonging to the FIEM is uniquely expressible as a linear combination, with non-negative integer coefficients, of those EM.

Example. The *be*-matrix B_{HCN} for H—C≡N: is expressed as follows in terms of its basis vectors.

$$B_{HCN} = \begin{pmatrix} 0 & 1 & 0 \\ 1 & 0 & 3 \\ 0 & 3 & 2 \end{pmatrix} \begin{matrix} H \\ C \\ N \end{matrix} = (P_{12} + P_{21}) + 3(P_{23} + P_{32}) + 2\,P_{33}$$

$$\phantom{B_{HCN} = } \begin{matrix} H & C & N \end{matrix}$$

2. Closed Shell *be*-Matrices

In closed shell chemistry, the *be*-matrices have even entries on the main diagonal. Making the

Definition. Let $CB(n)$ be the set of all $n \times n$ symmetric matrices with non-negative integer entries and with each diagonal entry even (or zero).

then each *be*-matrix describing an EM of closed shell chemistry belongs to $CB(n)$.

Clearly, $CB(n) \subset B(n) \subset S(n)$; and equally clearly $CB(n)$ is not a subgroup of $S(n)$. Letting $G[CB(n)]$ be the smallest subgroup of $S(n)$ containing. $CB(n)$, we have

Theorem 2. $G[CB(n)]$ is a proper subgroup of $S(n)$, with basis

$$\{P_{ij} + P_{ji} \mid 1 \le i < j \le n\} \cup \{2\,P_{ii} \mid i = 1,\ldots,n\}$$

Therefore $G[CB(n)]$ has rank $\dfrac{n(n+1)}{2}$ and consequently is isomorphic to $S(n)$.

41

Proof.   It is evident that the elements of the proposed basis are linearly independent. Let H be the subgroup of $S(n)$ having the indicated basis; it is clear that H is a proper subgroup of $S(n)$ because the diagonal entries of any $Q \in H$ are all even. Since any subgroup A containing $CB(n)$ must contain all the elements in the basis for H (because each one belongs to $CB(n)$, we find that A must contain H. Therefore $H = G[CB(n)]$ and we are done.

3. Determination of the Number of Molecules and Cyclic Structures from be-Matrices

An (i,j) cross in a matrix denotes a line through the ith row and j^{th} column of the matrix, $i.e.$, a vertical and a horizontal line through the (i,j)-entry of the matrix; (i,j) is called the vertex of the cross.

Let B be a given be-matrix. Starting with any non-zero entry b_{ij}, draw the (i,j)-cross. For each non-zero entry $(b_{i_1 j_1})$ on this cross, draw the (i_1,j_1)-cross, and continue this process of adding new crosses until each non-zero entry appearing on the resulting network is itself the vertex of a cross belonging to the network. Such a network is called a *grill*.

Theorem 3.   The atoms $i_1,\ldots,i_s$ crossed out by the horizontal (or vertical) lines in a grill are the atoms in a molecule of the EM represented by the be-matrix.

Proof.   The non-zero entries on the arms of the cross through (i,j) give all the atoms linked to atom i. The crosses on the non-zero entries encountered give further atoms linked to any atom joining i by either bonds or electron sharing. Continuing in this way, the grill represents mutually linked atoms, and none of those remaining are joined in any way to any atom on the grill. Thus, the $(i_1,\ldots,i_s)$ determine a molecule of the EM.

If the grill does not go through every non-zero entry of B, repeat this process starting at some uncovered element, to get an additional grill; with any entry not on these two grills, obtain another grill. Thus we get from Theorem 11 10 that

Corollary.   The total number of distinct grills on B is precisely the number of molecules in the EM represented by B.

In particular, we could use the grills to relabel the atoms of the EM so that the be-matrix appears in block-diagonal form, the number of blocks being the number of molecules in the EM. Observe that, if there are no rows/columns without non-zero entries, representing unbonded cores without free electrons, $e.g.$, $H^\oplus$, $Li^\oplus$, He, Ne, etc.) in that case, the number of rows/columns without non-zero entries will be on a separate grill.

This result shows that we can determine the number of molecules in an EM directly from the *be*-matrix representing that EM. We can, in addition, also determine whether or not the EM contains a molecule with cyclic bond structure directly from the *be*-matrix.

Theorem 4.   Let B be an $n \times n$ *be*-matrix for an EM. Let
(1) $k =$ number of molecules in the EM
(2) $b =$ the number of off-diagonal entries in B.
Then the EM contains a molecule with cyclic bond structure if and only if

$$n < k + \tfrac{1}{2} b$$

Proof.   We shall regard the EM as a 1-complex, K. The atoms are the vertices of K, and each bond or electron coupling is a 1-simplex of K. Thus, K has n vertices; and it has precisely $b/2$ 1-simplexes, because each pair of bonds (i,j), (j,i), $i \neq j$, determine a 1-simplex of K. Let $H_0(K)$ be the zero integral homology group of K; its rank $rH_0(K)$ is precisely the number of components in K; thus, in our case, $rH_0(K)$ is precisely the number of molecules in the EM.

Let $H_1(K)$ be the integral first homology group of K; its rank $rH_1(K)$ is the maximal number of linearly independent 1-cycles in K, so that $rH_1(K) > 0$ is a necessary and sufficient condition that there exists a molecule in the EM having cyclic bond structure.

According to the Euler-Poincaré formula [36], we have for K that (number of vertices) — (number of 1-simplexes) $= rH_0(K)$ $- rH_1(K)$. Thus, in our case,

$$n - b/2 = k - rH_1(K)$$

so

$$rH_1(K) = k - n + b/2$$

and the theorem is proved.

Remark 1.   Whenever it is known *a priori* that the EM represented by B has no molecule with cyclic structure, then one need not use grills to determine (since $rH_1(K) = 0$) that
$$k = n - b/2.$$

Remark 2.   The number $k - n + b/2$ is the maximal number of linearly independent 1-cycles. Depending on how one counts cycles, the actual number may differ from $rH_1(K)$: For example, consider the 1-complex determined by the vertices and edges of a tetrahedron; Euler-Poincaré gives $rH_1(K) = 3$, whereas some chemists may feel that there are four three-membered cycles and three four-membered cycles, and others that there is simply one.

4. Relation to Permutation Matrices

Recall that an $n \times n$ permutation matrix is a matrix having n entries 1 and all the rest zero, such that no two of the ones are in a common row, or common column. To relate be-matrices with permutation matrices, the main burden will be carried by a theorem of D. König:

Theorem 5. (König) [37] Suppose A is an $n \times n$ matrix of non-negative real numbers such that each row sum and each column sum is a fixed number $\sum \neq 0$. Then there exists in A a family of n non-zero entries, with no two of them being in the same row or column.

To apply this theorem, we require the notion of a "correction" for each be-matrix.

Let B be an $n \times n$ be-matrix. Since B is symmetric, it follows that for each $k = 1, \ldots, n$ the sum of the kth row is exactly the sum of the kth column, i.e.,

$$\sum_{i=1}^{n} b_{ik} = \sum_{j=1}^{n} b_{kj}$$

For each column k, let

$$b_k = \sum_{i=1}^{n} b_{ik} = \text{sum of entries in } k^{\text{th}} \text{ column}$$

$$b = \max_{k=1,..n} [b_k - b_{kk}]$$

$$= \text{largest column sum when diagonal terms are omitted}$$

We now remove each diagonal term b_{ii} and replace it by

$$\hat{b}_i = b - (b_i - b_{ii})$$

Clearly, $\hat{b}_i \geq 0$. In this new matrix $\hat{B}$, the sum of each column (and each row) is exactly $(b_i - b_{ii}) + \hat{b}_i = b$, so all row sums and all column sums, have the common value b, and all entries are non-negative integers.

Definition. Given $B \in B(n)$, the correction matrix $\hat{B}$ is
$$\hat{B} = B - \text{diag}(b_{11}, \ldots, b_{nn}) + \text{diag}(\hat{b}_1, \ldots, \hat{b}_n)$$
where $b_i = b - (b_i - b_{ii})$
and $b = \max[b_1 - b_{11}, \ldots, b_n - b_{nn}]$

We observe that
$$\hat{B} = B + \text{diag}\,(b-b_1, \ldots, b-b_n)$$
(For example, in the matrix M_{1+2} we have $b = 4$ and $\hat{M}_{1+2}$ has the same off diagonal terms, but the diagonal is diag $(3,0,3,2,3,0,1)$).

We also observe that $\hat{B}$ has at least one diagonal entry zero: in fact, if
$$b_k - b_{kk} = \max\,[b_1 - b_{11}, \ldots, b_n - b_{nn}]$$
then $\hat{b}_k = 0$.

Now, because of its definition, each row and each column of $\hat{B}$ has exactly the sum b, and all entries are non-negative integers. König's theorem applies to $\hat{B}$ so there is a system of n non-zero entries on distinct rows and columns. Let Q_1 be the permutation matrix with the n "ones" on those rows and columns, and consider $\hat{B} - Q_1$. This is a matrix with each row and each column sum exactly $b-1$, and with non-negative entries, so König's theorem applies to this new matrix also. Repeating the argument, after b steps we have
$$\hat{B} - Q_1 - \ldots - Q_b$$

is a matrix of non-negative terms, each row/column sum being zero. This matrix is therefore the zero matrix, and we have proved.

Theorem 6. Let B be an $n \times n$ be-matrix, let b_i be the ith row sum, and let $b = \max\,(b_1 - b_{11}, \ldots, b_n - b_{nn})$. Then
$$B = \sum_{i=1}^{b} Q_i - \text{diag}\,(b-b_1, \ldots, b-b_n)$$

where the Q_i are permutation matrices.

Thus, M_{1+2} is the sum of 4 permutation matrices — diag $(3,0,3,2,3,0,1)$.

Remark 1. Although $\hat{B}$ is a symmetric matrix, it is NOT true that the P_i can be chosen to be symmetric matrices.

Example.
$$\begin{pmatrix} 0 & 0 & 1 & 1 \\ 0 & 2 & 0 & 0 \\ 1 & 0 & 0 & 1 \\ 1 & 0 & 1 & 0 \end{pmatrix} = \begin{pmatrix} 0 & 0 & 0 & 1 \\ 0 & 1 & 0 & 0 \\ 1 & 0 & 0 & 0 \\ 0 & 0 & 1 & 0 \end{pmatrix} = \begin{pmatrix} 0 & 0 & 1 & 0 \\ 0 & 1 & 0 & 0 \\ 0 & 0 & 0 & 1 \\ 1 & 0 & 0 & 0 \end{pmatrix}$$; but it

is easy to see that one cannot choose ANY symmetric permutation matrix from the given matrix.

Remark 2. The representation of $\hat{B}$ (and therefore of B) as a sum of permutation matrices is NOT necessarily unique (even if one has all the matrices in the sums symmetric!).

$$Ex. \quad \begin{pmatrix} 1 & 0 & 0 & 1 \\ 0 & 1 & 1 & 0 \\ 0 & 1 & 1 & 0 \\ 1 & 0 & 0 & 1 \end{pmatrix} = \begin{pmatrix} 1 & 0 & 0 & 0 \\ 0 & 1 & 0 & 0 \\ 0 & 0 & 1 & 0 \\ 0 & 0 & 0 & 1 \end{pmatrix} + \begin{pmatrix} 0 & 0 & 0 & 1 \\ 0 & 0 & 1 & 0 \\ 0 & 1 & 0 & 0 \\ 1 & 0 & 0 & 0 \end{pmatrix} = \begin{pmatrix} 1 & 0 & 0 & 0 \\ 0 & 0 & 1 & 0 \\ 0 & 1 & 0 & 0 \\ 0 & 0 & 0 & 1 \end{pmatrix} + \begin{pmatrix} 0 & 0 & 0 & 1 \\ 0 & 1 & 0 & 0 \\ 0 & 0 & 1 & 0 \\ 1 & 0 & 0 & 0 \end{pmatrix}$$

Remark 3.   Since there are $n!$ permutation $n \times n$ matrices, the Theorem 3 leads to the following estimate: The set of $n \times n$ be-matrices in which max $[b_i - b_{ii}] = b$ is $\leq (n!)^b$.

C. Properties of Reaction Matrices

1. Integral Reaction Matrices

Let an FIEM be fixed, and let E, B be two EM belonging to the FIEM. A chemical reaction $B \to E$ is represented by a reaction matrix $R = E - B$. Formalizing this,

Definition.   Let B be the be-matrix for the starting materials, and E the be-matrix for the end products of a chemical reaction. The reaction matrix of $B \to E$ is the matrix $R = E - B$.

A reaction matrix is therefore a "make-break" indicator of bonds between pairs of atoms and of electron exchange between pairs of atoms, endowed with double book-keeping capability for bonds and electrons. The off-diagonal entries r_{ij} ($i \neq j$) indicate how many bonds between atom i and atom j are made ($+$) or broken ($-$) by the reaction; the diagonal entries r_{ii} reveal how many free electrons atom i gains ($+$) or loses ($-$) in the reaction $B \to E$.

Theorem 7.   A reaction matrix R for an FIEM with n atoms is always a symmetric $n \times n$ matrix with integer entries, and with sum over all entries equal to zero.

Proof.   The symmetry follows immediately from the equation $R = E - B$, as does the fact that the entries are $\pm$ integers or 0. Since the sum over all entries of B is the same as the sum over all entries of E (this amounts to saying that the bonds, and free electrons, in B are redistributed in E, or equivalently, that a necessary condition for E to be obtainable from B by some reaction is that the sum of the entries in E be equal to the sum of the entries in B), the second assertion is immediate.

Thus, if we make the

Definition.   Let $R(n)$ be the set of all $n \times n$ symmetric integral matrices with the sum over all entries equal to zero.

then the set of all reaction matrices corresponding to an interconversion of two EM belonging to an n-atom FIEM will belong to $R(n)$. The zero matrix, which belongs to $R(n)$, corresponds to no reaction.

It is clear that $R(n)$ is, in fact, a *subgroup* of $S(n)$- for, the difference of any two members of $R(n)$ is also a symmetric $n \times n$ integral matrix with sum over all terms zero.

Observe that no member of $R(n)$, other than the zero matrix, can belong to $B(n)$, since the sum of the entries in a $B \neq 0$ is necessarily >0; thus, $R(n) \cap B(n) = (0)$. In the geometric representation $\pi : B(n) \rightarrow R^s$, $\pi[B(n)]$ was visualized as a cone in R^s with vertex at the origin; we will see from Theorem 8 below that $\pi[R(n)]$ lies on a linear subspace going through the origin and having no other point in common with $\pi[R(n)]$:

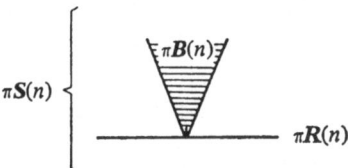

We will now produce a basis for the subgroup $R(n) \subset S(n)$.

Definition.   For each pair of integers (i,j) with $1 \leq i < j \leq n$, let

$$U_{ij} = P_{ii} - P_{jj}$$

This is an $n \times n$ matrix with i^{th} diagonal entry 1, the jth diagonal entry -1, and all the rest zero; it clearly belongs to $R(n)$.

Remark.   In chemical terms, an elementary *redox* process R1 is the exchange of electrons between pairs of atoms:

$$A_i\cdot \; + \; \cdot A_j \; \rightleftharpoons \; A_i \; + \; :A_j \tag{R1}$$

(for example, $Li\cdot + \cdot H \rightleftharpoons Li^{\oplus} + :H^{\ominus}$)

Thus, each reaction matrix U_{ij} represents an elementary redox reaction of type R1$\leftarrow$ in which an electron is transferred from atom j of the EM to atom i; $-U_{ij}$ corresponds to the reverse reaction. For this reason, we will call U_{ij} a redox reaction matrix.

Definition.   For each pair of integers (i,j) with $1 \leq i \leq j \leq n$, let

$$V_{ij} = P_{ij} + P_{ji} - P_{ii} - P_{jj}$$

This is an $n \times n$ matrix in which the diagonal entries v_{ii}, v_{jj} are both -1, the $v_{ij} = v_{ji} = +1$, and all the rest are zero; clearly, it belongs to $R(n)$.

Remark.   In chemical terms, an elementary homoaptic process R2 is bond making by electron sharing:

$$A_i\cdot \; + \; \cdot A_j \; \rightleftharpoons \; A_i \; - \; A_j \tag{R2}$$

(for example, $\cdot H + H\cdot \rightarrow H - H$, and its reverse is homolysis). Thus each reaction matrix V_{ij} represents an elementary homoaptic process

R2$_{\rightarrow}$, in which the atoms i and j each lose one free electron to form a bond; R2$_{\leftarrow}$ is the homolytic process. We will call V_{ij} a covalency matrix.

With the two center bond concept of chemical constitution, the redox and homoaptic processes are two elementary processes by which the redistribution of valence electrons in an FIEM takes place. The force of the following basis theorem is that *any* chemical reaction involving systems with integral bond orders is *uniquely* expressible as a combination of these two elementary processes:

Theorem 8. $\boldsymbol{R}(n)$ is a free abelian group of rank $\dfrac{n(n+1)}{2} - 1$, having

$$\{V_{ij} \,|\, 1 \leq i < j \leq n\} \cup \{U_{in} \,|\, i = 1,\ldots,n-1\}$$

as a basis. Moreover, the expression of any $R = (r_{ij})$ belonging to $\boldsymbol{R}(n)$ in terms of this basis is

$$R = \sum_{i<j} r_{ij} V_{ij} + \sum_{i=1}^{n-1} r_i U_{in}$$

where r_i is the ith row sum.

Proof.   First, we establish that the family $\{V_{ij}\} \cup \{U_{in}\}$ is linearly independent, i. e., if $\sum\limits_{i<j} \lambda_{ij} V_{ij} + \sum\limits_{i=1}^{n-1} \lambda_i U_{in} = 0$, then all λ_{ij} and all λ_i must be zero. Expressed directly in terms of the P_{ij}, the linear dependence states

$$\sum_{i<j} \lambda_{ii}[P_{ij} + P_{ij} - P_{ii} - P_{jj}] + \sum_{i=1}^{n-1} \lambda_i[P_{ii} - P_{nn}] = 0$$

Each $P_{i<j}$, occurs only once in this sum and with coefficient λ_{ij} so, upon collecting terms we have a linear combination

$$\sum_{i<j} \lambda_{ij}(P_{ij} + P_{ji}) + \sum \mu_k P_{kk} = 0 .$$

Since the family $\{P_{ij} + P_{ji}\} \cup \{P_{ii}\}$ is linearly independent, this requires that each $\lambda_{ij} = 0$. Thus, any linear dependence can involve only the terms U_{in}: $\sum\limits_{i=1}^{n-1} \lambda_i(P_{ii} - P_{nn}) = 0$. Noting that each P_{is} $i \neq n$, occurs only once in this summation, and with coefficient λ_i, we conclude that each $\lambda_i = 0$.

Since $\{V_{ij}\} \cup \{U_{in}\}$ is a linearly independent set, we need show only that each element of $\boldsymbol{R}(n)$ is expressible as a linear combination of the V_{ij} and U_{in} to assure that they form a basis. We do this in two steps.

(a) Let $R = (r_{ij})$ be any symmetric matrix. Then

$$
\sum_{i<j} r_{ij}V_{ij} = \begin{pmatrix} r_{11}-r_1 & r_{12}\cdots\cdots\cdots r_{1n} \\ r_{21} & r_{22}-r_2\cdots\cdots\cdots r_{2n} \\ -\;-\;-\;-\;-\;-\;-\;-\;-\;- \\ r_{n1} & r_{2n} \qquad\qquad r_{nn}-r_n \end{pmatrix}
$$

Indeed, $\sum\limits_{i<j} r_{ij}V_{ij} = \sum\limits_{i<j} r_{ij}(P_{ij}+P_{ji}) - \sum\limits_{i<j} r_{ij}(P_{ii}+P_{jj})$. The first sum on the right gives a symmetric matrix having exactly the same off-diagonal entries as R, since R is symmetric. To analyze the second summand, we introduce constants c_{ij} defined by

$$
\begin{aligned}
c_{ij} &= 1 \quad \text{if } i<j \\
&= 0 \quad \text{if } i\geq j
\end{aligned}
$$

so we can write

$$
\sum_{i<j} r_{ij}(P_{ii} + P_{jj}) = \sum_{i=1}^{n}\sum_{j=1}^{n} c_{ij}r_{ij}(P_{ii}+P_{jj})
$$

$$
= \sum_{i=1}^{n}\left[\sum_{j=1}^{n} c_{ij}r_{ij}\right] P_{ii} + \sum_{i=1}^{n}\left[\sum_{j=1}^{n} c_{li}r_{li}\right] P_{ii}
$$

$$
= \sum_{i=1}^{n}\left[\sum_{j>i} r_{ij} + \sum_{s<i} r_{si}\right] P_{ii}
$$

and, since $r_{si} = r_{is}$, this gives

$$
= \sum_{i=1}^{n}\left[\sum_{i\neq j} r_{ij}\right] P_{ii} = \sum_{i=1}^{n}\left[r_i-r_{ii}\right] P_{ii}
$$

and establishes (a).

Next, we establish

(b) Let $R = (r_{ij})$ be any matrix. Then

$$
\sum_{i=1}^{n-1} r_i U_{in} = \begin{pmatrix} r_1 \cdot & & 0 \\ & \cdot\;\cdot & \\ & \cdot\,r_{n-1} & \\ & & -\sum\limits_{1}^{n-1} r_i \\ 0 & & \end{pmatrix}
$$

For

$$
\sum_{i=1}^{n-1} r_i U_{in} = \sum_{i=1}^{n} r_i P_{ii} - \left(\sum_{i=1}^{n-1} r_i\right) P_{nn}
$$

It now follows from (a) and (b) that, if $R = (r_{ij})$ is any symmetric $n \times n$ matrix, then

$$\sum_{i<j} r_{ij}V_{ij} + \sum_{i=1}^{n-1} r_i U_{in} = \begin{pmatrix} r_{11} & \cdots\cdots\cdots & r_{1n} \\ r_{21} & \cdots\cdots\cdots & r_{2n} \\ & & \\ r_{n1}, & \cdots\cdots\cdots & r_{nn} - \sum_{1}^{n} r_i \end{pmatrix}$$

In the case of a reaction matrix, $\sum_{1}^{n} r_i = 0$, and the theorem is proved.

Remark 1. Given any family of numbers $\{\lambda_{ij} \mid 1 \le i < j \le n\}$, the method in the above proof shows that $\sum_{i<j} \lambda_{ij}V_{ij}$ is the reaction matrix obtained as follows: The upper off-diagonal entries are the λ_{ij}, the lower off-diagonal entries are the symmetrization (i.e., $\lambda_{ij} = \lambda_{ij}$ for each $i = j$) and each diagonal entry is the negative of the sum of all of the entries on that row.

Remark 2. Given any family $\{\lambda_i \mid i = 1, \ldots, n-1\}$ of numbers, $\sum_{1}^{n-1} \lambda_i U_{in}$ is the diagonal $n \times n$ matrix diag $\left[\lambda_1, \ldots, \lambda_{n-1}, -\sum_{1}^{n-1} \lambda_i\right]$

Remark 3. We have $S(n) = R(n) \oplus Z$; and the basis $\{V_{ij}\} \cup \{U_{in}\}$ for $R(n)$ enlarges to a basis for $S(n)$ by adding to it the matrix P_{nn}; with this basis, each element $A = (a_{ij})$ of $S(n)$ is uniquely expressible as

$$\sum_{i<j} a_{ij}V_{ij} + \sum_{i=1}^{n-1} a_i U_{in} + \left(\sum_{i=1}^{n} a_i\right) P_{nn}$$

This follows immediately from Remarks 1 and 2.

2. Closed Shell Reaction Matrices

In closed shell chemistry, all the elements of the set $CB(n)$ have each diagonal entry even, or zero. Thus, making the

Definition. Let $CR(n)$ be the set of all integral $n \times n$ symmetric matrices, with sum over all entries equal to zero, and with each diagonal entry even (or zero).

then every reaction matrix in closed shell chemistry will belong to $CR(n)$.
Clearly, $CR(n) \subset R(n)$; and, in fact, $CR(n)$ is a proper subgroup of $R(n)$.

We now prove

Theorem 9. $CR(n)$ is a free abelian group of rank $\dfrac{n(n+1)}{2} - 1$, having

$$\{V_{ij} + U_{ij} \,|\, 1 \leq i < j \leq n\} \cup \{2\,U_{in}\,|\, i=1,\ldots,n-1\}$$

as a basis. Moreover, the expression of any $R = (r_{ij})$ belonging to $CR(n)$ in terms of this basis is

$$R = \sum_{i<j} r_{ij}(V_{ij} + U_{ij}) + \sum_{i=1}^{n} \lambda_i\, 2\,U_{in}$$

where each $\lambda_{j_0} = \sum_{i<j_0} r_{ij_0} + \dfrac{1}{2} r_{j_0 j_0}$

Proof. We have, exactly, that

$$V_{ij} + U_{ij} = P_{ij} + P_{ji} - 2\,P_{jj}$$
$$2\,U_{in} = 2\,(P_{ii} - P_{nn})$$

so it is clear that each member of $\{U_{ij} + V_{ij}\}$ and of $\{2\,U_{in}\}$ belongs to $CR(n)$. The linear independence of the family $\{U_{ij} + V_{ij},\, 2\,U_{in}\}$ is proved exactly as in Theorem 4; and a calculation as in that theorem gives the above expression of a member of $CR(n)$ in terms of this basis.

Remark. The basis vectors $(V_{ij} + U_{ij})$ of closed shell chemistry correspond to the elementary *heteroaptic* and *heterolytic* reactions R2$_\rightarrow$ and R3$_\leftarrow$.

$$\text{R3:}\ A_i + \text{:}A_j \ \rightleftharpoons\ A_i - A_j$$

These are illustrated by this example:

$$:\overset{\cdot\cdot}{\underset{\cdot\cdot}{O}} + :N(CH_3)_3 \longrightarrow :\overset{\cdot\cdot}{\underset{\cdot\cdot}{O}} - N(CH_3)_3 \tag{R3}$$

A reaction matrix R of the interconversion H—C≡N: ⇌ H—N≡C: is expressed in terms of its basis vectors as follows:

$$R = \begin{pmatrix} 0 & -1 & +1 \\ -1 & +2 & 0 \\ +1 & 0 & -2 \end{pmatrix} = -(U_{12} + V_{12}) + (U_{13} + V_{13})$$

Because $R(n)$ and $CR(n)$ have the same rank, they are isomorphic. We exploit this to establish a relation between general and closed shell chemistry:

Theorem 10. Let $\Delta: S(n) \to S(n)$ be the homomorphism determined by

$$\Delta(P_{ij} + P_{ji}) = P_{ij} + P_{ji} + 2\,P_{ii} \qquad 1 \le i < j \le n$$

$$\Delta(P_{ii}) = 2\,P_{ii} \qquad\qquad\qquad i = 1,\ldots,n$$

Then Δ is an automorphism of $S(n)$ which simultaneously maps $B(n)$ INTO $CB(n)$ and $R(n)$ ONTO $CR(n)$.

Proof. We know that $\{P_{ij} + P_{ji},\ P_{ii}\}$ is a basis for $S(n)$; as in the proof of Theorem 4, we ascertain that the $\dfrac{n(n+1)}{2}$ elements $\{P_{ij} + P_{ji} + 2\,P_{ii},\ 2\,P_{ii}\}$ are linearly independent, consequently they, too, form a basis for $S(n)$. It follows then that Δ is an automorphism of $S(n)$.

It is evident that, if $B \in B(n)$, then $\Delta(B) \in CB(n)$. Now, consider the effect of this automorphism on the subgroup $R(n) \supset S(n)$: we have

$$\begin{aligned}
\Delta(V_{ij}) &= \Delta(P_{ij} + P_{ji} - P_{ii} - P_{jj}) = \Delta(P_{ij} + P_{ji}) \\
&\quad - \Delta(P_{ii} + P_{jj}) \\
&= P_{ij} + P_{ji} + 2\,P_{ii} - 2\,P_{ii} - 2\,P_{jj} = P_{ij} + \\
&\quad + P_{ji} - 2\,P_{jj} = V_{ij} + U_{ij}
\end{aligned}$$

and

$$\Delta(U_{in}) = \Delta(P_{ii} - P_{nn}) = 2\,(P_{ii} - P_{nn}) = 2\,U_{in}$$

Thus, Δ sends the basis for $R(n)$ bijectively onto the basis for $CR(n)$, consequently Δ maps $R(n)$ isomorphically ONTO $CR(n)$.

For each $A \in S(n)$, call $\Delta(A)$ its analog. This theorem states, essentially, that every relation in general chemistry has a unique analog in closed shell chemistry; given a reaction $E - B = R$ in general chemistry (i. e., $E,B \in B(n)$, $R \in R(n)$) we have $\Delta(E)$, $\Delta(B) \in CB(n)$, $\Delta(R) \in CR(n)$ and $\Delta(E) - \Delta(B) = \Delta(R)$ is the unique analogous reaction in closed shell chemistry.

3. Restricted Reaction Matrices

Let E,B be be-matrices for an FIEM belonging to restricted chemistry, and let $R = E - B$ be the reaction matrix for $B \to E$. In accordance with our definition of restricted chemistry, we have for each atom i of the FIEM that $e_{ii} = b_{ii}$ and that the row sum b_i in B is equal to the row sum e_i in E.

Consequently, a reaction matrix in restricted chemistry will have all diagonal entries zero and each row sum equal to zero. Thus, if we make the

Definition.   Let $R(n)$ be the set of all $n \times n$ symmetric integral matrices with all diagonal entries zero and each row sum equal to zero.

then the set of all reaction matrices in the restricted chemistry of an n-atom FIEM is contained in $R(n)$.

Clearly, $R(n) \subset CR(n) \subset R(n) \subset S(n)$; and, equally clearly, $R(n)$ is itself a subgroup of $S(n)$.

Theorem 11. $R(n)$ is a free abelian group of rank $\dfrac{n(n-3)}{2}$.

Proof.   We identify $S(n)$ with its geometrical representation in Z^s, $s = \dfrac{n(n+1)}{2}$, and denote the elements of Z^s by $(x_{11}, \ldots, x_{1n}; x_{22}, \ldots, x_{2n}; \ldots,; x_{nn})$. For each pair of integers (i,j), $1 \leq i \leq j \leq n$, let p_{ij} be the element of Z^s with (i,j) coordinate 1 and the rest zero; then $\{p_{ij} \mid 1 \leq i \leq j \leq n\}$ is a basis for Z^s. In each (i,j), let $h_{ij}: Z^s \to Z$ be the homorphism determined by $h_{ij}(p_{kl}) = \delta_{ik} \cdot \delta_{jl}$.

Recall that, for any free abelian group A and homomorphism h: $A \to Z$, we find from the formula, rank (Image h) + rank (Kernel h) = rank A, that if h is not identically zero, then rank (Kernel h) = (rank A) − 1.

Now consider the $2n$ homomorphisms $H_l: Z^s \to Z$, where

$$H = h_{ll} \qquad l = 1, \ldots, n$$

$$H_{n+l} = \sum_{i=1}^{n} h_{il} \qquad l = 1, \ldots, n$$

Since we have identified $R(n)$ with $\pi R(n)$, we note that $\pi R(n)$ is exactly the intersection $\bigcap\limits_{i=1}^{2n}$ Ker H_i. Since for each $1 \leq k \leq 2n$, no H_k is identically zero on $\bigcap\limits_{i=1}^{k-1}$ Ker H_i, each Ker H_i reduces the rank of $S(n)$ by 1, consequently,

$$\text{rank } \pi R(n) = \frac{n(n+1)}{2} - 2n = \frac{n(n-3)}{2}$$

and the proof is complete.

This says that there are essentially $\dfrac{n(n-3)}{2}$ distinct basic reactions in the restricted chemistry of an n-atom FIEM; all the others are obtained from these by linear combination. We shall now produce a basis for $R(n)$.

Corollary. For each pair (i,j) of integers, $1 \leq i < j \leq n\text{-}2$, let $L(i,j) = V_{ij} + V_{n-1,n} - V_{i,n-1} - V_{j,n}$, and for each $i = 1, \ldots, n\text{-}3$, let $K(i) = V_{i,n-1} + V_{i+1,n} - V_{i,n} - V_{i+1,n-1}$. Then $\{L(i,j) \mid 1 \leq i < j \leq n\text{-}2\} \cup \{K(i) \mid i = 1, \ldots, n\text{-}3\}$ is a basis for $\boldsymbol{R}(n)$.

Proof. It is clear that each $L(i,j)$ and $K(i)$ belongs to $\boldsymbol{R}(n)$; and there are a total of $\dfrac{n(n-3)}{2}$ elements in $\{L(i,j)\} \cup \{K(i)\}$. It is therefore enough to prove the system linearly independent, in order to assure it constitutes a basis.

Assume, then, that $\displaystyle\sum_{1 < i < j < n-2} \lambda_{ij} L(i,j) + \sum_{1}^{n-3} \lambda_i K(i) = 0;$ we want to show that the λ_{ij}, λ_i are all zero. Expressed directly in terms of the V_{ij}, this linear dependence states that

$$\sum_{1 < i < j < n-2} \lambda_{ij}[V_{ij} + V_{n-1,n} - V_{i,n-1} - V_{jn}] +$$

$$+ \sum_{i}^{n-3} \lambda_i[V_{i,n-1} + V_{i+1,n} - V_{i,n} - V_{i+1,n-1}] = 0$$

Each V_{ij}, $h \leq i < j < n\text{-}2$ occurs exactly once in this summation and with coefficient λ_{ij} so, upon collecting terms, we have a linear dependence

$$\sum_{1 < i < j < n-2} \lambda_{ij} V_{ij} + \text{(linear combination of remaining terms)}$$
$$= 0.$$

Since we know that the set $\{V_{ij} \mid 1 \leq i < j \leq n\}$ is linearly independent, this requires that each $\lambda_{ij} = 0$.

Thus, any linear dependence can involve only the terms $K(i)$:

$$\sum_{i=1}^{n-3} \lambda_i[V_{i,n-1} + V_{i+1,n} - V_{i,n} - V_{i+1,n-1}] = 0 .$$

Upon collecting terms, we note that $V_{1,n}$ occurs only once, with coefficient λ_1 so that $\lambda_1 = 0$. In the surviving terms, V_{2n} occurs only once, with coefficient λ_2, so that λ_2 must be zero. Proceeding recursively, we find $\lambda_3 = \ldots = \lambda_{n-3} = 0$. Thus $\{L(i,j), K(i)\}$ is a linearly independent set, and the proof is complete.

Remark. The representation of a given $R \in R(n)$ in terms of this basis is not so simple as in the previous cases; although the coefficient of each $L(i,j)$ will be r_{ij}, the coefficients of the $K(i)$ are more complicated, being sums of suitably chosen entries in the matrix R.

There are some further algebraic properties of $R(n)$ that are simple to establish.

Theorem 12. (a) There is no non-trivial restricted reaction matrix for $n \leq 3$.

(b) Any non-zero restricted reaction matrix must have at least four non-zero rows.

Proof. (a) Clearly, $R(1)$ consists only of zero, and it is trivial to verify that $R(2)$ does also. For $n = 3$, we get rank $R(3) = 0$ from Theorem 9, which says the only member of $R(3)$ is the zero matrix.

(b) This can be established directly from the basis, but also as follows: observe that if $A \in R(n)$ has a zero row, then because it is symmetric the corresponding column is also zero. Removing that row and column would then give an $(n-1) \times (n-1)$ restricted reaction matrix. Thus, if A had only $k < 4$ non-zero rows, we could, by successively removing the zero rows and corresponding columns, obtain a $k \times k$ non-zero reaction matrix. But, since $k < 4$, this is impossible by part (a).

We also have

Theorem 13. For each $R \in R(n)$

(a) The sum of the terms $\sum_{i<j} r_{ij}$ in the upper triangle is zero.

(b) Det $R = 0$.

Proof. (a) Since each row sum $\sum_{j=1}^{n} r_{ij} = 0$, we find

$$0 = \sum_{i,j} r_{ij} = \sum_{i<j} r_{ij} + \sum_{i>j} r_{ij}$$

But
$r_{ii} = 0$ and $r_{ij} = r_{ji}$, so $0 = 2 \sum_{i<j} r_{ij}$, which gives (a).

(b) Regarding each row $(r_{i1}, \ldots, r_{in})$ as a vector in R^n, we have $\sum_{i=1}^{n} (r_{i1}, \ldots, r_{in}) = (0, \ldots, 0)$, which says that the row vectors are linearly dependent. As is well-known, this is equivalent to the statement that the determinant of R is zero.

4. Relation to Permutation Matrices

Let $B \to E$ have the $n \times n$ reaction matrix R. According to Theorem 4, we have

$$B = \sum_1^b P_i - \mathrm{diag}\,(b-b_1, \ldots, b-b_n)$$

$$E = \sum_1^e Q_i - \mathrm{diag}\,(e-e_1, \ldots, e-e_n)$$

where P_i, Q_i are $n \times n$ permutation matrices. Since $R = E - B$, we obtain from Theorem 6:

Theorem 14. Let R be the reaction matrix of $B \to E$. Let the ith row sums of B resp E be b_i resp e_i and let

$$b = \max\,[b_1-b_{11}, \ldots, b_n-b_{nn}],$$
$$e = \max\,[e_1-e_{11}, \ldots, e_n-e_{nn}].$$

Then

$$R = \left(\sum_1^e Q_i - \sum_1^b P_i\right) - \mathrm{diag}\,[(e-e_1) - (b-b_1), \ldots,$$

$$(e-e_n) - (b-b_n)]$$

where the Q_i, P_i are permutation matrices.

Remark.   If $e = b$, then R is (aside from a diagonal matrix) a sum of e differences of permutation matrices.

D. Fitting Reaction Matrices

Definition.   Let R be a fixed reaction matrix. R is said to fit a matrix B if $B + R$ is a be-matrix.

This means, in chemical terms, that R represents a reaction that can be applied to the EM represented by B. In geometrical terms, the definition says that if B is a member of the cone $\boldsymbol{B}(n)$, and R is a reaction matrix, then R fits B if $R+B$ also belongs to the cone $\boldsymbol{B}(n)$.

Using the notation $A \geq 0$ to denote that all the entries of A are non-negative, the definition of "fitting" can be formalized by

Theorem 15. R fits B if and only if $R+B \geq 0$.

Proof.   The "if" alone requires proof; Since R,B are symmetric, so also is $R+B$; since all entries are non-negative integers, the matrix $R+B$ belongs to the cone $\boldsymbol{B}(n)$.

Now, fix a be-matrix B, and consider the set of all possible reaction matrices that fit B, i. e.,

Definition.   For any given $B \in \boldsymbol{B}(n)$, let $\boldsymbol{R}(B) = \{\boldsymbol{R} \mid R+B \geq 0\}$.
We shall investigate the set $\boldsymbol{R}(B)$. Our first result is that the knowledge of the set $\boldsymbol{R}(B)$ for *one* $B \in \boldsymbol{B}(n)$ immediately gives us the totality of all reaction matrices fitting any other EM that can be obtained from B by some reaction, precisely

Theorem 16. Let $E,B \in \boldsymbol{B}(n)$, and let $E = B + R_0$. Then a reaction matrix Q fits E if and only if $Q = R - R_0$ for some R fitting B. Thus,

$$\boldsymbol{R}(E) = \{R - R_0 \mid R \in R(B)\} \ .$$

Proof.   Suppose Q fits E; then $B + R_0 + Q = E + Q \geq 0$; letting $R = Q + R_0$, this says R fits B and $Q = R_0$. Conversely, if $Q = R - R_0$ for some R fitting B, then $E + Q = (B + R_0) + (R - R_0) = B + R \geq 0$ so Q fits E.

Remark.   An alternative formulation of this theorem is: R fits $B + R_0$ if and only if $R + R_0$ fits B.

We now study the fitting of components of reaction matrices. Specifically, suppose B is a be-matrix and R a reaction matrix fitting B. Represent R as a sum of two simpler reaction matrices, $R = R_1 + R_2$. It may very well happen that R_1 fits B, but R_2 does NOT fit B; or also that neither R_1 nor R_2 alone fit B.

Example. Let $B = \begin{pmatrix} 0 & 0 & 1 \\ 0 & 1 & 1 \\ 1 & 1 & 0 \end{pmatrix}$ and $R = \begin{pmatrix} 0 & 1 & -1 \\ 1 & -1 & 0 \\ -1 & 0 & 1 \end{pmatrix}$. Now write R as the sum $R = R_1 + R_2$ of two homolytic-homoaptic reactions,

$$R_1 = \begin{pmatrix} 1 & 0 & -1 \\ 0 & 0 & 0 \\ -1 & 0 & 1 \end{pmatrix}, \qquad R_2 = \begin{pmatrix} -1 & 1 & 0 \\ 1 & -1 & 0 \\ 0 & 0 & 0 \end{pmatrix}$$

It is clear that R_1 fits B, but that R_2 does *not* fit B. As a trivial example of the possibility that neither summand fits, write the zero reaction matrix as diag $[-2,2,0]$ + diag $[2,-2,0]$; this sum of course fits B, but neither summand fits B.

Using the geometric interpretation $\pi\colon S(n) \to Z^s$, the reason for this non-commutative behaviour is easily understood: Writing $R = R_1 + R_2$, the vector R_2 may be too "large" so that $B + R_2$ is no longer in the cone $B(n)$, although R_1 returns $B + R_2$ back into the cone.

We have seen that each reaction matrix R can be represented uniquely as a sum $\sum_{i<j} r_{ij}V_{ij} + \sum_{1}^{n=1} r_i U_{in}$. To investigate the fitting properties of the summands, we need the

Lemma.   Let $B \geq 0$ be a *be*-matrix and let $R = \sum_{i<j} r_{ij}V_{ij} + \sum_{i=1}^{n-1} r_i U_{in}$
fit B, so that $B + R \geq 0$. Then

(a) If any $r_{ij} < 0$, then $r_{ij}V_{ij}$ fits B.

(b) If all $r_{ij} \geq 0$, but some $r_i < 0$, then $r_i U_{in}$ fits B.

(c) If all $r_{ij} \geq 0$, and all $r_i > 0$, then each $r_i U_{in}$ fits B.

(d) If all $r_{ij} \geq 0$, and all $r_i = 0$, then each $r_{ij}V_{ij}$ fits B.

Proof.   (a) Assume $r_{ij} < 0$, and form $B + r_{ij}V_{ij}$. The (ij) entry in this sum is $b_{ij} + r_{ij}$, which is non-negative because $B + R \geq 0$; the ith diagonal entry is $b_{ii} - r_{ij} = b_{ii} + |r_{ij}| \geq 0$, and similarly the jth diagonal entry is non-negative. The only entries affected being the (ij), (ji), (ii), (jj), this shows $r_{ij}V_{ij}$ fits B.

(b) We have $r_i U_{in} = \text{diag}[0, \ldots, r_i, \ldots, -r_i]$. Thus, $b_{nn} - r_1 = b_{nn} + |r_1| \geq 0$. Now, consider $b_{ii} + r_i$; we know $b_{ii} + r_{ii} \geq 0$ by hypothesis; moreover, in the matrix R, we know $r_{ii} = r_i - (\sum_{j \neq i} r_{ij})$. Therefore,

$$b_{ii} = r_i - (\sum_{j \neq i} r_{ij}) = b_{ii} + r_{ii} \geq 0$$

so

$$b_{ii} + r_i \geq \sum_{j \neq i} r_{ij} \quad \text{and, because all } r_{ij} \geq 0, \text{ the latter sum}$$
is ≥ 0.

(c) In this case, we have $b_{ii} + r_i \geq 0$. Now, we know that the (n,n) entry of the matrix R is $-\sum_{1}^{n-1} r_i - (r_{n1} + \ldots + r_{n,n-1})$; because R is fitting, we have

$$b_{nn} - \sum_{1}^{n-1} r_i - \sum_{i=1}^{n-1} r_{ni} \geq 0$$

This says

$$b_{nn} - r_i \geq \sum_{j \neq i} r_i + \sum_{i=1}^{n-1} r_{nk} \geq 0$$

because all terms of the sums are non-negative.

(d) We consider $B + r_{ij} V_{ij}$ because $r_{ij} \geq 0$, the off-diagonal entries $b_{ij} + r_{ij}$, $b_{ij} + r_{ji}$ are both non-negative.

Now, since all the $r_i = 0$, the ith diagonal term in R is $-\sum_{\substack{j \\ i \neq j}} r_{ij}$ and, because of the fitting, we know

$$b_{ii} - \left(\sum_{i \neq j} {}^{jr}ij \right) \geq 0$$

This says

$$b_{ii} - r_{ij} - \sum_{\substack{k \\ k \neq i,j}} rt_k \geq 0$$

or,

$$b_{ii} - r_{ij} \geq 0$$

This leads to

Theorem 17. Let B be a be-matrix, and let $R = \sum_{i<j} r_{ij} V_{ij} + \sum_{i=1}^{n-1} r_i U_{in}$ fit B.

(1) Let $R_1, \ldots, R_a$ be the elementary homolytic terms with $r_{ij}, < 0$.

(2) Let $R_{a+1}, \ldots R_b$ be the elementary redox terms with $r_i < 0$.

(3) Let $R_{b+1}, \ldots, R_c$ be the elementary redox terms with $r_i > 0$.

(4) Let $R_{c+1}, \ldots, R_d$ be the elementary homolytic terms with $r_{ij} > 0$.

Then, for each integer $1 \leq q \leq d$, we have

$$B + R_1 + \ldots + R_q \geq 0.$$

Alternatively stated: When R is written *in the order* $R_1 + \ldots + R_a + \ldots + R_b + \ldots + R_c + \ldots + R_d$, then each $R_1 + \ldots R_q$ fits B; or, in other terms, R_1 fits B, R_2 fits $B + R_1$, R_3 fits $B + R_1 + R_2$, etc. In chemical terms, we first break suitable bonds, and then make new ones.

Proof.   We have $B \geq 0$ and $B + R \geq 0$. Observe that, for each q,

$$R_{q+1} + \ldots + R_d \quad \text{fits} \quad B + R_1 + \ldots + R_q$$

because $(B + \ldots + R_q) + (R_{q+1} + \ldots + R_q) = B + R \geq 0$. Thus, if we define

$$B_q = B + R_1 + \ldots + R_q,$$

we have, for each q, that $R_{q+1} + \ldots + R_d$ fits B_q.

We also observe, because $\{V_{ij}, U_{in}\}$ is a basis for $\boldsymbol{R}(n)$, that the unique representation of $R_{q+1}, + \ldots + R_d$ is exactly $R - (R_1 + \ldots + R_q)$, i. e., that obtained from R by removing the r_{ij}, r_i used in the first q of the R_i. We are now ready to use the Lemma. According to the Lemma, part (a), we have $B_1 = B + R_1 \geq 0$; according to our observation, $R_2 + \ldots + R_d$ fits B_1. Thus, we can use the Lemma, part (a), again, applied this time to B_1 and $R_2 + \ldots + R_d$ to find $B_2 = B_1 + R_2 \geq 0$. We proceed stepwise until all $r_{ij}, < 0$ are used, to find $B_3 \geq 0, \ldots, B_a \geq 0$. Now, $R_{a+1} + \ldots R_d$ fits B_a, and no longer contains any $r_{ij}, < 0$, so we start using part (b) of the Lemma, ascertaining stepwise that $B_{a+1} \geq 0, \ldots, B_b \geq 0$. At this stage, we use part (c) of the Lemma, because $R_{b+1} + \ldots + R_d$ has no $r_{ij}, < 0$ and no $r_i < 0$, to find $B_{b+1} \geq 0, \ldots, B_c \geq 0$. Since $R_{c+1} + \ldots + R_d$ has only terms $r_{ij} V_{ij}$, and all $r_{ij} \geq 0$, part (d) fo the Lemma applies to give $B_{c+1} \geq 0, \ldots, B_d \geq 0$ and the proof is complete.

This result remains true in closed shell chemistry. Specifically

Theorem 18. Let B be a closed shell *be*-matrix and let $R = \sum_{i<j} r_{ij}$ $(V_{ij} + U_{ij}) + \sum_{i=1}^{n-1} i(2U_{in})$ be a closed shell reaction matrix fitting B.

(a) Let $R_1, \ldots, R_a$ be the terms $r_{ij}(V_{ij}+U_{ij})$ with $r_{i,} < 0$.
(b) Let $R_{a+1}, \ldots, R_b$ be the terms $\lambda_i(2U_{in})$ with $\lambda_i < 0$.
(c) Let $R_{b+1}, \ldots, R_c$ be the terms $\lambda_i(2U_{in})$ with $\lambda_i > 0$.
(d) Let $R_{c+1}, \ldots, R_d$ be the terms $r_{ij}(V_{ij}+U_{ij})$ with $r_{ij} > 0$.

Then for each integer $1 \leq q \leq d$, we have

$$B + R_1 + \ldots + R_q \geq 0,$$

and it is a closed shell matrix.

Proof.

The proof is identical to that of Theorem 16, once the appropriate version of the Lemma is established. For that purpose, one need only verify that in each of the cases (a)—(d) of the Lemma, the numbers entering the diagonal are always even; and verification of this is straightforward.

Remark.

Suppose R is a *restricted* reaction matrix. Then R has three representations: in terms of the basis for $R(n)$, for $CR(n)$ or for $R(n)$. As theorem 16 shows, the elementary homolytic and redox components of R, when added to B in proper order, will give a sequence of be-matrices at each stage; however, since each elementary operation affects the diagonal, very few (if any) of these intermediary B_i will belong to restricted chemistry; and the same is true if one uses the basis for $CR(n)$.

Even more important, however, is that if a restricted reaction matrix is decomposed according to the basis $L(ij)$, $K(i)$, then it may happen that, although R fits B, *none* of the $L(ij)$, $K(i)$, or any linear combination omitting one of the terms, fits B. For consider in $R(5)$ the matrix.

$$2\,L(1,2) = \begin{pmatrix} 0 & 2 & 0 & 0 & -2 \\ 2 & 0 & 0 & -2 & 0 \\ 0 & 0 & 0 & 0 & 0 \\ 0 & -2 & 0 & 0 & 2 \\ -2 & 0 & 0 & 2 & 0 \end{pmatrix}$$

$$-2\,L(2,3) = \begin{pmatrix} 0 & 0 & 0 & 0 & 0 \\ 0 & 0 & -2 & 0 & 2 \\ 0 & -2 & 0 & 2 & 0 \\ 0 & 0 & 2 & 0 & -2 \\ 0 & 2 & 0 & -2 & 0 \end{pmatrix}$$

$$-3\,K(1) = \begin{pmatrix} 0 & 0 & 0 & -3 & 3 \\ 0 & 0 & 0 & 3 & -3 \\ 0 & 0 & 0 & 0 & 0 \\ -3 & 3 & 0 & 0 & 0 \\ 3 & -3 & 0 & 0 & 0 \end{pmatrix}$$

J. Dugundji and I. Ugi

having sum

$$R = \begin{pmatrix} 0 & 2 & 0 & -3 & 1 \\ 2 & 0 & -2 & 1 & -1 \\ 0 & -2 & 0 & 2 & 0 \\ -3 & 1 & 2 & 0 & 0 \\ 1 & -1 & 0 & 0 & 0 \end{pmatrix}$$

and let

$$B = \begin{pmatrix} 0 & 0 & 1 & 3 & 0 \\ 0 & 0 & 3 & 0 & 1 \\ 1 & 3 & 0 & 0 & 0 \\ 3 & 0 & 0 & 0 & 0 \\ 0 & 1 & 0 & 0 & 0 \end{pmatrix}$$

Then R fits B, but none of $2\,L(1,2)\,-2\,L(2,3)$, $-3\,K(1)$, or the sum of any two of them, fits B; moreover, neither $L(1,2)$, $L(2,3)$ nor $K(1)$ fits B.

This example shows *no* result such as Theorems 16 and 17 is true in restricted chemistry, at least in terms of the basis for $R(n)$ that we have selected.

Acknowledgements. The authors each gratefully acknowledge partial support of the National Science Foundation, b   grants # GP 28927X (I. U.) and # GP (J. D.).

References and Abbreviations

1) ACM = Atom Connectivity Matrix; *be*-matrix = *bond* and *electron* matrix, also *beginning* ($= B$) and *end* ($= E$) matrix of chemical reactions; EM = Ensemble of Molecules; FIEM = Family of Isomeric Ensembles of Molecules; CRT = Cathode Ray Tube.

2) Rollette, J. S., ed.: Computing methods in crystallography. London: Pergamon 1965. — Stout, G. H., Jensen, L. H.: X-Ray structure determination, a practical guide. London: Macmillan 1970. — Krüger, C.: Angew. Chem. *84*, 412 (1972); Angew. Chem. Intern. Ed. Engl. *11*, 387 (1972).

3) Dewar, M. J. S.: Molecular orbital theory of organic chemistry. New York: McGraw-Hill Book Co. 1969. — Bünau, G. V.: Angew. Chem. *84*, 418 (1972); Angew. Chem. Intern. Ed. Engl. *11*, 393 (1972), and references cited therein.

4) Review articles, see *e. g.*:
Kresze, G.: Angew. Chem. *82*, 563 (1970); Angew. Chem. Intern. Ed. Engl. *9*, 545, (1970). — Meyer, E.: Angew. Chem. *82*, 563 (1970); Angew. Chem. Internat. Edit. Engl. *9*, 545 (1970). — Fugmann, R., Nickelsen, H., Winter, J. H.: Angew. Chem. *82*, 611 (1970); Angew. Chem. Intern. Ed. Engl. *9*, 589 (1970).

5) Lederberg, J.: The mathematical sciences, p. 37. Cambridge, Mass.: M. I. T. Press 1969. — Buchs, A., Duffield, A. M., Schroll, G., Djerassi, C., Delfino, A. B., Buchanan, B. G., Sutherland, G. L., Feigenbaum, E. A., Lederberg, J.: J. Am. Chem. Soc. *92*, 6831 (1970). — Henneberg, D., Casper, K., Zeigler, E., Weimann, B.: Angew. Chem. *84*, 381 (1972); Angew. Chem. Intern. Ed. Engl. *11*, 357 (1972).

[6] Review articles: Wilke, G.: Angew. Chem. *84*, 370 (1972); Angew. Chem. Intern Edit. Engl. *11*, 347 (1972). — Zeigler, E., Henneberg, D., Schomburg, G.: Angew. Chem. *84*, 371 (1972); Angew. Chem. Intern. Ed. Engl. *11*, 348 (1972). — Schomburg, G., Weeke, F., Weimann, B., Ziegler, E.: Angew. Chem. *84*, 391 (1972); Angew. Chem. Intern. Ed. Engl. *11*, 366 (1972). — Hoffmann, E. G., Stumpfle, W., Schroth, G., Weimann, B., Ziegler, E., Brandt, J.: Angew. Chem. *84*, 400 (1972); Angew. Chem. Intern. Ed. Engl. *11*, 375 (1972), and literature cited therein.

[7] Flores, I.: Computer programming. Englewood Cliffs: Prentice-Hall 1966. — Detar, D. T.: Computer programs for chemists. New York: Benjamin 1968. — Kalzan, H.: Advanced programming. New York: Van Nostrand Reinhold 1970.

[8] Minsky, M.: Computers and thought (eds. E. Feigenbaum and J. Feldman), p. 406. New York: McGraw-Hill 1963. — Minsky, M.: Computation: Finite and infinite machines, Englewood Cliffs: Prentice-Hall 1967. — Minsky, M., ed.: Semantic information processing, Cambridge, Mass.: M. I. T. Press 1968. — Nilsson, N. J.: Problem solving methods in artificial intelligence, New York: McGraw-Hill 1971.

[9] Corey, E. J.: Pure Appl. Chem. *14*, 19 (1967). — Corey, E. J., Wipke, W. T.: Science *166*, 178 (1969). — Corey, E. J., Wipke, W. T., Cramer, R. D., Howe, W. J.: J. Am. Chem. Soc. *94*, 421, 431 (1972). — Corey, E. J., Cramer, R. D., Howe, W. J.: J. Am. Chem. Soc. *94*, 440 (1972). — Corey, E. J., Peterson, G. A.: J. Am. Chem. Soc. *94*, 460 (1972).

[10] Hendrickson, J. B.: J. Am. Chem. Soc. *23*, 6847 (1971).

[11] Stevens, R. V.: Personal communication.

[12] Ugi, I., Kaufhold, G.: (unpubl. 1966—67); Review articles: Ugi, I.: Rec. Chem. Progr. *30*, 289 (1969); Intra-Science Chem. Rep. *5*, 229 (1971). — Gokel, G., Hoffmann, P., Kleimann, H., Klusacek, H., Ludke, G., Marquarding, D., Ugi, I.: In: Isonitrile chemistry (ed. I. Ugi), p. 201. New York: Academic Press 1971.

[13] Gelernter, H.: Topics Curr. Chem. *41*. Submitted 1973.

[14] Nachr. Chem. Techn. *20*, 147 (1972) (Interview on Synthetic Design with R. B. Woodward and A. Eschenmoser).

[15] Finlay, A.: A hundred jears of chemistry, p. 157. New York: McMillan 1937.

[16] Nachr. Chem. Techn. *20*, 127 (1972).

[17] Dugundji, J., Gillespie, P. D., Ugi, U., Marquarding, D.: In: Chemical applications of graph theory (ed. A. Balaban). New York: Academic Press (in press), and literature cited therein.

[18] Bellman, R. E., Dreyfus, S.: Applied dynamic programming, Princeton: Princeton University Press 1962; see also: Colomb, S., Baumert, L.: J. A. C. M. *12*, 516 (1965). — Dreyfus, S.: Operations Res. *17*, 395 (1969).

[19] Ugi, I., Kaufhold, D.: Liebigs Ann. Chem. *709*, 11 (1967).

[20] Preuss, H.: Intern. J. Quant. Chem. *3*, 123, 131 (1969), and references contained therein.

[21] Ugi, I., Marquarding, D., Klusacek, H., Gokel, G., Gillespie, P.: Angew. Chem. *82*, 741 (1970); Angew. Chem. Intern. Ed. Engl. *9*, 703 (1970).

[22] Ege, G.: Naturwissenschaften *58*, 247 (1971).

[23] Ugi, I.: Jahrb. 1964 Akad. Wiss., p. 21. Göttingen: Vandenhoeck and Rupprecht 1965; Z. Naturforsch. *20B*, 405 (1965). — Ruch, E., Ugi, I.: Theoret. Chim. Acta *4*, 287 (1966); Top. Stereochem. *4*, 99 (1969). Ruch, E.: Accounts Chem. Res. *5*, 49 (1972).

[24] Polya, G.: Acta Math. *68*, 145 (1937). — Golomb, S. W.: Information Theory (Fourth London Symposium), p. 404. London: Butterworths, 1961. — De Bruijn, N. G.: In: Applied combinatorial mathematics (ed. F. F. Beckenbach), p. 140. New York: Wiley 1964. and preceding communications. — Ruch, E., Hässelbarth, W., Richter, B.: Theoret. Chim. Acta *19*, 288 (1970).

25) Ugi, I., Marquarding, D., Klusacek, H., Gillespie, P., Ramirez, F.: Accounts Chem. Res. *4*, 288 (1971). — Gillespie, P. D., Hoffmann, P., Klusacek, H., Marquarding, D., Pfohl, S., Ramirez, F., Tsolis, E. A., Ugi, S.: Angew. Chem. *83*, 691 (1971); Angew. Chem. Internat. Edit. *10*, (1971). — Gillespie, P., Marquarding, D., Ramirez, F., Ugi, I.: Angew. Chem. *85*, *99* (1973); Angew. Chem. Intern. Edit. Engl. *12*, *91* (1973).

26) Ugi, I., Gillespie, P. D.: Angew. Chem. *83*, 493 (1971); Angew. Chem. Intern. Ed. Engl. *10*, 503 (1971). —

27) Ugi, I.: Intra-Science Chem. Rep *5*, 229 (1971). — Ugi, I., Gillespie, P.: Angew. Chem. *83*, 982, 990 (1971); Angew. Chem. Intern. Ed. Engl. *10*, 914, 915 (1971). — Ugi, I., Gillespie, P. D., Gillespie, C.: Trans. N. Y. Acad. Sci. *34*, 416 (1972).

28) See *e. g.*, the fragment strategies of peptide syntheses:
Bodanszky, M., Ondetti, M. A.: Peptide synthesis. New York: Wiley (Interscience) 1966.

29) Spialter, L.: J. Am. Chem. Soc. *85*, 2012 (1963); J. Chem. Doc. *4*, 261, 269 (1964).

30) Lipscomb, G. N.: Boron hydrides. New York: Benjamin 1963.

31) Lewis, G. N.: J. Am. Chem. Soc. *38*, 762 (1916). — Langmuir, I.: J. Am. Chem. Soc. *41*, 868, 1543 (1919).

32) See *e. g.*, March, J.: Advance organic chemistry, reactions, mechanisms, and structure. New York: McGraw-Hill 1968, and references cited therein.

33) Kuratowski, K.: Topologie. P. A. N. monog. math., polish scientific publishers, Warszawa, Vol. I, 1958; Vol. II, 1961. — Kowalski, J.: Topological spaces. New York: Academic Press 1965. — Dugundji, J.: Topology. Boston: Allyn and Bacon 1966. — Kuratowski, K., Mostowski, A.: Set theory. Amsterdam: North Holland Publ. Co. 1968.

34) Prelog, V.: Abstract of the Roger Adams Award Lecture on June 17, 1969, at the ACS Meeting, Salt Lake City, Utah. — Prelog, V., Helmchen, G.: Helv. Chim. Acta *55*, 2581 (1972).

35) Hamermesh, M.: Group theory and its applications to physical problems. Reading, Mass.: Addison-Wesley 1962. — Knödel, W.: Graphentheoretische Methoden und ihre Anwendung. Berlin-Heidelberg-New York: Springer 1969.

36) Harary, F.: Graph theory, Reading, Mass.: Addison-Wesley 1969.

37) König, D.: Theorie der endlichen und unendlichen Graphen. Leipzig 1936; reprinted New York: Chelsea 1950.

Received July 27, 1972

Computer Techniques for Retrieval of Information from the Chemical Literature

Dr. Douglas C. Veal

Assistant Director, United Kingdom Chemical Information Service, University of Nottingham, Nottingham, England

Contents

I. The Structure of the Literature

Scientists need information for a variety of reasons — to keep up-to-date on their current projects, to assemble information for new projects, to determine specific facts, to stimulate ideas, and so on. They obtain the information they need by a variety of means — reading the literature, attending conferences, discussing with colleagues, etc. Of the available means, various studies [1,2] have shown that the primary and secondary literature sources are the most important. The reasons for this are not hard to find. Firstly, private communications, "invisible colleges", conferences, etc., are all to a greater or lesser extent restricted, whereas the published literature is available to all. Secondly, published papers provide the established written record of results. As Cahn [3] has pointed out "their value is immediate but also permanent; they contain the essential facts and arguments by means of which science progresses, the solid foundation and the testing ground".

The basic repository of the information is the primary literature. The vast majority of information occurs in the journal literature although patents and reports also play a significant part. The size of this primary literature presents severe problems to scientists attempting to use it to find items of interest to them. For example, Chemical Abstracts Service currently scan more than 13,000 journals of chemical interest, and select some 300,000 articles per year. This body of literature is growing at a rate of about 9% per annum compounded [4,5]. There have long been expectations that this rate would slow down, but so far there is no evidence of any abatement [6]. Faced with this volume of literature, it appears that most chemists find time to read only 5—10 primary journals regularly [7,8], whereas there is evidence [9] that for most chemists items of interest occur in at least 20 journals, even over a relatively short period of time. For fairly broad topics such as "Electrochemistry", Cahn (Ref. 3, Appendix XI) showed that relevant references occurred in well over 100 different journals in the course of a year.

It is not surprising, therefore, that there has long been a substantial secondary literature in chemistry to help users to access the primary literature. Basically, this secondary literature performs two functions. Firstly, by gathering together, classifying, sorting and indexing primary articles, it helps users to locate items of potential interest, especially in journals which they would not otherwise see. Secondly, by preparing a condensed version of the original article, such as an abstract, or even just a title, it helps users to decide which items are in fact of interest without the chore of scanning large numbers of original articles.

The secondary literature thus exists basically as a guide to the primary, although it has been reported [10] to be a useful source of information in itself. This, however, is more likely to be an indication of the inertia of users than of the information content of the secondary sources, since the compression involved in producing the latter (see Section III) inevitably means that only a very small proportion of the original information can be included. The secondary sources should therefore be considered to be indicative, rather than informative.

In addition to the primary and secondary sources, there is a growing body of review literature in which information on particular topics is gathered together, sifted, evaluated and reported. As the volume of literature continues to grow it is likely that these reviews and monographs will play an increasingly important role. It is also fairly common for these review-type articles to be rounded out with some original information.

We thus have the general pattern of a large and growing body of primary literature making necessary the existence of comprehensive secondary services, and a growing body of review literature. This is the basic structure of the literature from which information may be retrieved.

Note. It is regrettable, but true, that the literature of information retrieval suffers from all the defects of rapid growth, etc. to at least as great an extent as any other discipline. Fortunately, in the last few years a number of extremely useful review articles have appeared on key topics. It is inevitable that an article of such broad scope as the present one can only deal with matters superficially and I have therefore referenced this article extensively, so that those who wish to follow up particular interests may do so.

II. Basic Computer Techniques

The physical form of the literature is that of which this article is comprised *i.e.* an arrangement of characters on a printed page. The bulk of the material is in the form of running text, although there may also be diagrams, structural formulae, graphs, etc. A substantial proportion of the literature, particularly the primary literature, is available only in hard-copy form. However, more and more publishers are using computer-based production systems and so an increasing amount of information is becoming available in machine-readable form, and computer searching is becoming increasingly important.

Generally speaking, the computer-produced data bases consist of continuous strings of characters paralleling the running text of the printed versions. We thus see that information in both the primary and secondary literature is largely stored in the form of character strings. If we are searching for information we must therefore look for particular character strings.

When a chemist is searching manually his eye will scan a sequence of words which his brain will interpret. From his inherent knowledge of the subject he will from time to time recognise particular words or phrases as being relevant to the topic which interests him, but he will not usually formulate a list of significant words and phrases in advance. The computer, however, has no inherent knowledge of the topic. Therefore, a list of the significant "search terms" must be specified in advance and the computer will simply search for the specified sequences of characters. Thus, the essential process of computer searching is that of *character matching*.

Consider the search term "BENZENE". In the simplest case the computer will identify the first character of this as B. It will then look at the first character of the first word on the data base. If this is not B the terms cannot match and the computer will go on to the next term from the data base. When it finds a term beginning with a B it will check to see if the next character is an E, and so on until the match is definitely established or eliminated. However, this basic character-matching process has been subjected to many modifications and improvements. Computer searches are still most commonly carried out as batch processes (that is; a whole group of search terms is searched together against a data base) although on-line searching is now coming to the fore. In a batch search there will be a list of required search terms and a list of terms occurring in the data base. The search process consists of identifying matches between the two lists. Generally, these lists will be in a random order initially, *i.e.* the list of search terms may be in the order in which they were received, and the list of data terms in document order. To ease the problem of matching, it is logical to put one or other list (or both) into some more suitable order. For example, if one list is put into alphabetic order, then terms from the other list can be dealt with one by one, each in turn being sought in the ordered list. The ordering of the list means that the computer can confine its search to a small part of the list. Taking the above example, all the terms beginning with B will be gathered together. The computer can then rapidly scan down the list to the appropriate place to see whether there are any matching terms. For example, consider the list of terms shown in Fig. 1.

The computer begins by just matching the first letter, and rapidly scans down the list until it reaches BACON. From here it scans the second letter until it reaches BEACON, and so on until it reaches the matching term BENZENE. After this no terms can match and so the search is complete. This is obviously much quicker than having to match all search terms against all data-base terms.

Although both the search terms and the data-base terms can be put into alphabetic order, normally only one is in fact done. This process is called "inversion" and the ordered file is known as an "inverted file". The other list is then processed serially, from beginning to end, during the matching process, and is therefore known as a "serial file".

Obviously, two types of system can be created — that in which the search terms are ordered, and that in which the data-base terms are ordered. These are commonly referred to by the state of data-base file so the first is known

Fig. 1. Alphabetised list of terms

as *serial searching* and the second as *inverted-file searching*. Naturally, the aim of searching a literature file is to retrieve documents and in an inverted file search the terms must therefore carry the identifiers of the documents to which they pertain. Similarly, in a serial search, each search term must carry an identifier to relate it to the enquiry to which it belongs.

It is useful to consider file organisation in relation to clustering theory. Basically, the aim of clustering techniques is to recognise the characteristics which exist in a file and to use them to bring together file items which have something in common. In the context of information retrieval the file items will be documents, and their attributes will be index terms or their equivalents. In this sense we can see that a serial search file is a particular form of cluster, in which each cluster contains only one document, represented by its particular set of index terms. An inverted file is similarly a clustered file in which each cluster is represented by a single term, but may consist of one or more documents. Obviously these are two very simple file structures whereas more sophisticated clustering techniques lead to more complex file structures. Since these have been almost exclusively applied to textual information retrieval they are considered in more detail in Section V.

A document collection on which no specific file organisation has been imposed can be considered to be random, and the various clustering techniques are methods of imposing varying degrees of order upon the collection, so as to help searchers to locate material they require. This situation has clear parallels with thermodynamics which have recently been elegantly expostulated by Fugmann [11]. One of the most important conclusions which Fugmann draws is that, to maintain a given level of search capability, the degree of order must increase as the size of the document collection increases.

One of the problems of computer searching is that chemists may be interested not only in whole words, but also in fragments of words. For example, a chemist may be interested in all names which contain the fragment "PYRID". This facility is referred to as "term truncation" and is usually indicated by an asterisk, *e.g.* *PYRID*. In the above example, where the computer was searching for the complete word BENZENE, any word not beginning with a B could immediately be eliminated. Obviously, this is not the case when a word fragment is being sought since the fragment may start or finish anywhere within a word. Term truncation therefore inevitably adds substantially to the amount of character matching which must be performed. Nevertheless, it is such a valuable facility, particularly in chemistry, that most search systems allow for it.

The sort of character matching we have dealt with so far is linear and unidirectional, that is, characters are matched in sequence from left to right. This is applicable to textual and numeric information but multidimensional structural information obviously presents much greater difficulties. For example, a chemist may be interested in the following sequence of atoms and bonds:

$$\begin{array}{c} \overset{7\;8}{OH} \\ \overset{1\;\;2\;\;\;3\;\;\;4\;\;\;\overset{6}{|}\;\;9}{C\!=\!\!=\!\!C\!\!-\!\!C\!\!-\!\!} \\ 5 \end{array}$$

Here he is interested in a fragment of structure, rather than a word fragment, and this facility is known as "substructure searching". In substructure searching the computer may begin by searching through a file of structures for a carbon atom. It must then see if the next symbol is a double bond and so on. However, a non-match at a particular point does not necessarily rule out a structure. Consider the case in which the following structure is being searched to see if it contains the above fragment, with symbols being matched in the order shown:

$$-\overset{6}{O}\overset{5}{-}\overset{4}{C}\overset{3}{=}\overset{1}{C}\overset{2}{-}N-C$$

with branches labeled 7, 11, 10, 9, $HO-C8$, and 12.

The search will proceed as shown in Table 1.

Table 1

Step. No.	Sought	Found	Action
1	C	C	Match, look for next character
2	=	—	Non-match, return to C and try other direction
3	=	=	Match, look for next character
4	C	C	Match, look for next character
5	—	—	Match, look for next character
6	C	O	Non-match, go back to branch point and try other branch
7	—	—	Match, look for next character
8	C	C	Match, look for next character
9	—	—	Match, look for next character
10	O	O	Match, look for next character
11	H	H	Match, look for next character
12	—	—	Structure found, register a hit

This simple example illustrates the complexities of character matching structural information, in which path tracing may proceed in any direction and must cope with rings, branch points, etc.

So far we have talked about searching by matching one character at a time. Obviously, this process is well suited to those computers which are "byte-oriented" *i.e.* they handle information in units of one byte or one character (usually 6 or 8 binary "bits".) Conversely, "word-oriented" computers, which handle information in units of computer words (perhaps 36 or 48 bits) are less well adapted to character-matching routines. However, all computers can match fields larger than one character, and the greater the amount of information which can be matched simultaneously the more efficient will be the search. In the above discussion we talked in terms of matching one character at a time. By matching, say, two characters at a time, it would obviously be possible to establish identity or non-identity in fewer steps. However, such a simple technique has severe disadvantages for dealing with such things as term truncation (see above), so more elaborate techniques must be used.

A much more efficient application of matching larger units of data is the technique of screening. A screen is a certain number of bits set aside to record information about a particular file item. In its simplest implementation each bit relates specifically to one characteristic, being set to 1 if the item possesses the characteristic and left at 0 if it does not. For example, in a file of words, each word may have a screen of 26 bits, each bit corresponding to one letter of the alphabet. The bit screen for the word REVIEW would then look as follows:

```
A B C D E F G H I J K L M N O P Q R S T U V W X Y Z
0 0 0 0 1 0 0 0 1 0 0 0 0 0 0 0 0 1 0 0 0 1 1 0 0 0
```

In this case each word would have a screen of the same length — 26 bits. Since fixed-length records are easier to handle on computers this fact also helps to make screen searching very efficient. In searching, an equivalent screen is set up for the search term and then matched with all the screens of the search file. In each case the whole screen is matched as a unit as opposed to the sequential operation of character matching. Thus the search is essentially a one-step operation. The disadvantage of screening is that it does not necessarily give perfect results. For example, the above screen system would not distinguish anagrams. The word VIEWER would also have the above bit screen and would also register as a hit in screen search. For this reason screen searching is usually carried out as a preliminary to a more detailed character-by-character search. The idea is that the screen search will rapidly eliminate (or 'screen out') a large number of items that cannot possibly be hits (in our example, words which do not at least contain one each of the letters R, E, V, I, W) so that the more expensive character matching has only to be done on a small proportion of the total file.

The above example illustrates screening at its simplest. In practice, screens are more often multiples of 8 bits, to accord with normal computer practice; they may be open-ended (*i.e.* of variable length) rather than fixed length, and may be combined or superimposed in various ways. The technique. of superimposed coding was first developed by Mooers [12] for a purely mechanical punched card system, but has since been incorporated into a highly sophisticated computer system [13].

One of the problems with all information except numeric data is that the size of the units (words, chemical structures) is variable, and can be very large. The following peptide name, for example, has about 1000 characters, and the longest chemical name so far recorded has 1913 [14].

N-Carboxyglycyl-L-3-[1-(carboxymethyl)pyridinium-3-yl]-
L-alanyl-3-[1-(carboxymethyl)pyridinium-3-yl]-L-alanyl-
3-[1-(carboxymethyl)pyridinium-3-yl]-L-alanyl-3-[1-(carboxymethyl)pyridinium-3-yl]-L-alanyl-3-[1-(carboxymethyl)
pyridinium-3-yl]-L-alanyl-3-[1-(carboxymethyl)pyridinium
-3-yl]-L-alanyl-3-[1-(carboxymethyl)pyridinium-3-yl]-L-
alanyl-3-[1-(carboxymethyl)pyridinium-3-yl]-L-alanyl-3-
[1-(carboxymethyl)pyridinium-3-yl]-L-alanyl-3-[1-(carboxymethyl)pyridinium-3-yl]-L-alanyl-3-[1-(carboxymethyl)
pyridinium-3-yl]-L-alanyl-3-[1-(carboxymethyl)pyridinium
-3-yl]-L-alanyl-3-[1-(carboxymethyl)pyridinium-3-yl]-L-
alanyl-3-[1-(carboxymethyl)pyridinium-3-yl]-L-alanyl-3-
[1-(carboxymethyl)pyridinium-3-yl]-L-alanyl-3-[1-(carboxymethyl)pyridinium-3-yl]-L-alanyl-3-[1-(carboxymethyl)
pyridinium-3-yl]-L-alanyl-3-[1-(carboxymethyl)pyridinium-
3-yl]-L-alanyl-3-[1-(carboxymethyl)pyridinium-3-yl]-L-
alanyl-3-[1-(carboxymethyl)pyridinium-3-yl]-L-alanyl-3-
[1-(carboxymethyl)pyridinium-3-yl]-L-alanylglycine.

There is therefore considerable scope for the use of file compression techniques, both to compress the amount of data to be handled and to achieve conversion to fixed-length records.

Screening is of course one way of achieving this. If, for example, we are using a computer in which each character takes 8 bits, the word REVIEW will take 48 bits and the above name some 8000, whereas, as we have seen, a screen may take only 26 bits. However, screens only achieve overall file compression if they are used *instead of* full files, whereas they are generally used *in addition*.

There are standard methods of data compression which can be applied, arising out of fundamental information theory, that is, the minimum number of bits required to store any given amount of information unambiguously and without redundancy. The basic theory of the information content of a message was first propounded by Shannon in his classic paper [15], and file compression techniques arising out of that have since found many applications, for example, in the searching of files of spectral data [16]. One of the most commonly used techniques is "hash coding". In this, records of variable, and often considerable, length are reduced to relatively short fixed-length records by very simple manipulations. For example, a 16-bit record may be reduced to 8 bits by simple "adding together" the first 8 and second 8. Hash codes are thus really only another form of screen. An example of their use in information retrieval is given in Ref.[17].

Numeric information is already quite compact and so is less susceptible to data compression. In fact one of the commonest techniques for file compression of other types of data is to turn them into numbers. This is accomplished by operating some sort of registry or dictionary which records all the items encountered and assigns a unique number to each. As we shall see later, this has been applied to both words and chemical structures.

In all our discussion of computer searching so far, we have really only been concerned with establishing within a search file the existence of certain specified character strings. However, if we go back to our chemist doing his manual searching we will realise that it is not merely the occurrence of particular key words which "triggers him off"; he also takes into account their context. For instance he may be very interested in the ultraviolet spectra of unsaturated ketones but not at all interested in their infrared spectra. The mere occurrence of suitable chemical names will not alone be enough to arouse his interest.

This aspect, too, must be catered for in computer searching. Thus, it is not enough for a searcher to specify a list of required terms; he must also specify a logical schema into which the terms fit. This can be done in various ways, for example by the use of explicit Boolean logical statements, or by assigning weights to each search term and specifying a threshold weight which an item must "score" before it is retrieved. Current search systems allow very sophisticated logical statements to be built up. Users may specify not only that two terms occur in the same document but that they occur in a particular order, in the same sentence, and so on [18]. Not all searchers require such sophisticated facilities, but in a minority of cases they are of crucial value [19].

III. Types of Information

If we look at a piece of chemical text, we see that it is made up of characters, generally numeric digits, alphabetic characters, punctuation marks, and perhaps a few special symbols and diagrams. Information is conveyed by arranging these characters and symbols into particular sequences, words, numbers, etc., which convey particular meanings. Even a superficial study is sufficient to show that there are basically three different types of information — numerical, structural, and conceptual.

(i) Numerical information is essentially the record of quantitive experimental results such as physical properties, kinetic data, and so on. Such information generally takes the form of numbers, either in tables or incorporated in the running text, but it may also be recorded graphically.

(ii) Since chemistry is the study of chemical compounds, information, about the structure of the compounds being studied is crucially important. It has been estimated that "85% of the newly reported information in chemistry deals with chemical substances or compounds" [20]. Thus, structural information plays an important part in the majority of papers. It may occur as text, or "pseudo-text" (nomenclature, notations, line formulae), but most commonly takes the form of structural diagrams, particularly where reaction schemes are being discussed. These diagrams are readily understandable, and it is difficult to see how the information in the scheme shown in Fig. 2, for example, could be adequately conveyed just using the written word.

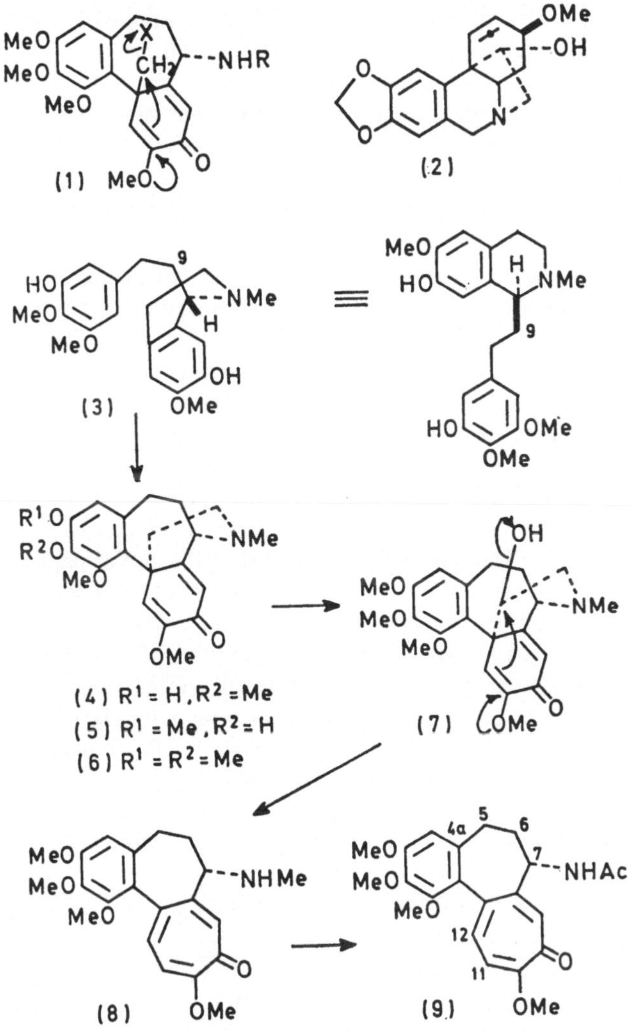

Fig. 2. Battersby, A. R., *et al.*: Reproduced from Alkaloid biosynthesis. Part XVIII. Biosynthesis of colchicine from the 1-Phenethylisoquinoline system. J. C. S. Perkin Trans. *I*, 1741 (1972), by kind permission of the Editor

(iii) Conceptual information is more difficult to define. Perhaps the best way of approaching it is to say that if the numerical information comprises the numbers in the paper and structural information comprises the diagrams, then conceptual information comprises the text. It embodies the essential ideas of the paper, and embodies them in the form of words since this is the medium we use to express ideas.

It should be obvious that this is a brief and simplified treatment of a very complex topic, and that the lines between the different kinds of information are by no means as distinct as is implied in the foregoing discussion. For example, all kinds of information can be, and often are, recorded as written text: structural information in such phrases as "the dihydroxy steroid" and quantitative information in such phrases as "an energy barrier of about four kcal mol^{-1}". Nevertheless, it is important to realise that there are different kinds of information since there are fundamental differences between them, some of which have important consequences for information retrieval. These differences have been discussed in detail elsewhere [21], but are summarised in Table 2. It is clear from the foregoing that each type of information presents its own problems for information retrieval. These will be discussed in more detail later.

It has already been pointed out that the retrieval procedures rely heavily on the secondary literature, rather than the primary. Apart from the differences between the various kinds of information discussed above, there are also variations in their treatment by the secondary services. It has been reported [3,20] that the average chemical paper consists of some 3400 words, made up to 5—6 pages with the inclusion of numerical tables, structural diagrams etc. There is evidence [22,23] that the size of papers is increasing. The average abstract, however, is only 3—4% of the size of the original. so a very substantial compression takes place. Large structural diagrams such as that shown in Fig. 2, graphs and tables of numeric data do not lend themselves to compression. Structural diagrams are usually included in abstracts only to the minimum extent necessary for clarity, what I have called pseudo-text being used wherever possible. For numeric data, only key results or a few representative figures are normally included. Conceptual information is more amenable to compression, but even here the secondary sources are rarely able to do more than give an indication of the paper's content. This presents us with problems in discussing information retrieval. The standard secondary abstracting and indexing services are really only suitable for alerting users to the existence of potentially interesting papers. For example, a user interested in thermodynamic data will search the secondary services for the *idea* of "thermodynamic data". This will alert him to papers likely to contain such data, but to get the actual data he has to refer to the original articles. We thus have the situation where the standard secondary services are used for *document* retrieval, and the primary documents are used for *information* retrieval. This is a distinction which is all too often allowed to become blurred.

In this article we are concerned with the use of computer techniques. There has been speculation that in due course the whole of the primary literature will become available in machine-readable form. Although some experiments are being done [24], the difficulties of achieving this, particularly for structural, graphical and tabular data, has meant that progress has been slow. So although computerized information retrieval direct from the primary literature may come in the future it is probably some years away. Current activity has been concerned with computerisation of the document retrieval function, using machine-readable secondary data bases.

Table 2

Type of information	Characteristics	Required search facilities
Numerical	Linear Small units (*i. e.* even large numbers have few digits) Small character set (10 digits)	Exact or inexact match (*e. g.* range searching)
Conceptual (including "pseudo-textual" structural information)	Linear Units may be large (*i. e.* words have many characters) Fairly large character set (usually around 100 characters)	Exact match Whole or part unit match (*i. e.* term truncation) with unidirectional path tracing
Structural	Three-dimensional Large units (molecules) Large character set (over 100 different atom and bond symbols)	Exact or inexact match Whole or part unit match (*i. e.* substructure search) with multidirectional path-tracing

However, retrieval services cater for both current-awareness and retrospective search needs. Whilst computerised document retrieval services are useful for current awareness, there has been much activity to apply computer techniques to actual *information* retrieval, particularly for retrospective searching. This has taken the form of setting up specialised data banks for numeric and structural information. These are a form of secondary service but are more closely related to handbooks and encyclopaedias than to abstract publications. (Data handbooks, etc., are sometimes referred to as tertiary sources, but since secondary sources are guides to the primary, the term "tertiary sources" is more accurately applied to guides to the secondary literature [25].)

We thus have the situation in which essentially three different kinds of computerised retrieval system have evolved. There are specialised systems dealing with structural and numeric data, with the emphasis on retrospective searching. There are also standard retrieval systems, using text searching, which are used for both current-awareness and retrospective searching, and which also cover structural and numeric data as far as document retrieval is concerned. We can now consider in more detail how each type of information fits into this pattern.

IV. Numeric Information

The treatment of numeric information is currently an area of substantial and rapid development. Activity in this field has been comprehensively reviewed [26] by *Codata*, a Committee set up by the International Council of Scientific Unions to coordinate developments. This review covers both computerised and non-computerised systems, but Codata also has a task group dealing specifically with automated information handling and this has also recently reported on the state of the art [27].

The content of the report emphasises that the boundaries between the different types of information *system* are just as ill-defined as those between the different kinds of information. Many of the so-called data banks consist of files of bibliographic references, and in the majority of cases the final output from a search consists of a list of references rather than raw data. Thus, the report concludes that "the goal of data center automation, computer storage of the data and their retrieval in response to requests for information, is not yet widespread". In other words, we are still in the era of *document* retrieval rather than *information* retrieval.

In order to see just what progress has been achieved it is worth considering the general characteristics of numeric data banks. First we must remember that information systems have two facets — the data bank that forms the basic file of the system, and some form or multiple forms of organisation (*i.e.* indexing and/or classification) to enable the file to be accessed. In a system concerned with numeric data, the data may constitute either the basic file or the method of indexing. In the first case, since the data in question are most commonly chemical or physical properties of compounds, the file is generally organised on the basis of chemical structure. This enables a user to find, for example, the properties of a particular compound, or class of compounds. (The latter case implies some sort of structural classification or substructure search facility, and demonstrates a very important point, namely the close interdependence of structure, text and numeric search techniques. This point should not be obscured by the fact that the three are here being dealt with separately.)

Examples of systems employing structural classifications are the *U. K. Mass Spectrometry Data Centre* [28], which uses a fairly general chemical classification; the *Sadtler Spectral Service* [29], which uses a classification based on elements and functional groups; and the *Environmental Mutagens Information Centre* at Oak Ridge in the U.S.A. [30] which uses the Wiswesser Line Notation or WLN [31].

We have already seen that numeric data banks exist primarily for retrospective searching. What is therefore required in those where numeric data consitute the basic file, is to record the most reliable values in the system. In other words a critical evaluation is really required, prior to input to the system. The promotion of such evaluations is an area where Codata is particularly active. Furthermore, particular values may be determined repetitively and one wishes to record only the "best" value. This may mean replacing a value with a later, more accurate one, so the facility to do this must be built into the system. Finally, primary data gathered from the literature may be transformed in some way or used to calculate secondary data which can also be stored. Examples are the *Information Centre for Mineral Thermodynamics* in Grenoble [32], which uses primary thermodynamic data to calculate other thermodynamic functions, and the *On-line Data Bank on Atomic and Molecular Physics* at Belfast [33], which is working on the automatic transformation and development of relations between different data sets. This last points the way towards exciting future developments in dynamic, as opposed to static, data banks.

The evaluation, maintenance and manipulative functions discussed above are peculiar to numeric data. Continuous updating, in particular, is very easy to carry out on computer files but is very difficult with the conventional printed data handbooks. We can therefore expect the former to gradually replace the latter over the next few years, provided only that access to computerised data banks can be made as readily and cheaply available.

One of the problems in performing the evaluative function has been the inconsistency with which numeric data are presented in the primary literature. It is encouraging that steps are now being taken to overcome this (*e.g.* [34,35]).

In the second case, that is where the basic file is of documents organised on the basis of numeric data, the problems are somewhat different. The data in question are still most commonly the properties of compounds, and the most common function is where as user is seeking to use known properties to identify a compound, or where, by identifying a group of compounds with a common property, he seeks structure/property relationships. The basic file is therefore generally of structures, as for example the Sadtler spectral file mentioned above [29]. However, in some cases the file may be of bibliographic data. The Mass Spectrometry Data Centre [28], for example, has such a file in addition to its structural file.

This brings us up against the problems of *searching* numeric data — in particular searching for values within a defined range, and the inexact matching of such things as spectral data, where the peaks are unlikely to coincide exactly in either wavelength or intensity. Details of this type of activity are being discussed elsewhere. However, numbers are very suitable for manipulation by computer, being compact, readily compared, etc. and suitable search techniques can be devised provided that the match criteria are clearly specified.

Certainly, the compact nature of numeric information allows very large volumes of data to be handled. For example, the CSISRS file of neutron cross-section data at the U.S. Oak Ridge National Laboratory contains over 1,000,000 data points [32].

V. Textual Information

As we have seen, in considering textual information we are concerned with information stored in the form of words. By and large this means conceptual information, although quantitative and structural data may also be included. In the normal course of events the information is written down in whatever words seem most appropriate to the author, abstractor, etc.: in other words the language used is "natural" or "uncontrolled". This means that a retrieval system based on natural language has to deal with potentially very large numbers of index terms. Various studies of vocabulary accretion rates [36,37] have shown that the vocabulary size does not level out as the file size grows, but rather tends to settle down to a constant growth rate. This rate appears to be related to the size of the discipline involved, and can be quite substantial for a broad area such as "chemistry". The formidable tasks of searching natural language, particularly before the advent of computers, led to the development of alternative approaches. For example, the Medlars system [38] uses a rigidly structured and controlled indexing vocabulary. However, even here, the breadth of the discipline being covered and the depth and specificity of indexing required means that the indexing vocabulary is quite large, 10,000 terms in the case of Medlars. The continuing growth of knowledge, and of the natural vocabulary used in the primary literature, means that the use of a controlled indexing language implies a considerable and continuous monitoring and updating effort. Also the "thermodynamic" considerations mentioned earlier [11] mean that, as the file size increases, the indexing must become ever more specific to retain the same level of effectiveness. However, the advent of computers has meant that other solutions to the problems of searching natural language can be sought. For example, term truncation can greatly reduce the number of search terms required; the use of screening techniques may increase the search efficiency [39,40]; and procedures for the automatic construction of search profiles [19,41] may obviate the need for users to perform laborious compilations of search term lists.

Much of the work in this area has arisen from the conversion to machine processing of formerly manual operations by the publishers of secondary literature. Taking Chemical Abstracts Service as a prime example, they began to computerise their operations in the early 1950's, primarily to increase their efficiency in coping with the ever-increasing flood of primary literature. As a result, their secondary data bases have begun to appear in computer-readable form, in essentially natural language, beginning with Chemical Titles in 1962. These data bases can then form the basis of computerised retrieval services such as those offered by UKCIS [42].

Discussion about the relative merits of free text versus controlled language frequently generates the fervour of a religious war and I do not want to enter into the debate here. Suffice it to say that no evidence has so far been produced to demonstrate that one approach has clear advantages over the other [43].

Apart from the use of vocabulary control, there are many systems which aid retrieval by classification. (Indeed, vocabulary control is only one form of classification). It has already been mentioned that the secondary services commonly classify and rearrange the documents they cover. These classifications are frequently computerised and so can be used for computer retrieval. For example, the Universal Decimal Classification [44] is in use as a basis of computerised systems (e. g. [45]). Much work is also going on in the development of automatic classifications, based on cluster analysis techniques [46,47]. In these cases the classification may be applied by assigning the appropriate codes to file items stored in some entirely different order (e. g. chronologically or alphabetically by source journal). Alternatively, the classification scheme may be used as a basis for organising the file, so that all file items falling into a particular class occur together. This is most commonly achieved by the use of the inverted file approach discussed in Section II.

We saw there that an inverted file consists of an ordered list of index terms, each term accompanied by the list of documents to which it pertains. It is thus nothing more than a computerised version of the customary printed index. It is the natural technique to use in controlled-language systems such as Medlars [38], but interestingly can equally well be applied to natural language [48]. The value of the inverted file approach increases as the average number of documents per index term grows. It can greatly increase the efficiency of searching although with the penalty of having to do a file inversion at some stage. Its value thus increases as the size of the search file increases and it is most useful for retrospective search of large files. Also, since profiles are processed serially, search term by search term, it is particularly suitable for on-line searching.

The inverted file is, then, merely a computer application of traditional indexing philosophy. Perhaps the most important difference between computer searching and manual search of a printed index is that the computer enables coordination to be done much more easily at search time. In a computer search, the required term coordinations can be specified, as a Boolean expression for example, much more precisely than could be achieved by any precoordinated indexing system. Computer systems therefore convey a much greater flexibility.

Although they are only indirectly concerned with information retrieval from the literature, it is worth noting that the computer has also made significant contributions to the production of printed indexes. Much of the work has involved the production of traditional indexes by computer methods, ranging from simple uniterm and keyword indexes to fully articulated subject indexes and dual dictionaries. Products more particularly of the computer age are permuted indexes such as the Permuterm Index® of the Institute for Scientific Information and the KWIC, KWAC, KWOC family.

One other significant contribution is in the field of citation indexing. Citation indexing was in fact first used towards the end of the last century but it passed out of use because, with the growth of the literature the manual methods in use could no longer cope with the vast amount of permutation and sorting involved. However, one of the things computers can do is to process large amounts of data very quickly and so computerised citation indexing is viable. It is in fact very successfully used, for example, by ISI in their *Science Citation Index* [49]. To illustrate the scale of operation required, this involves processing some 4,000,000 citations per year, from some 2000 source journals.

We have seen that, whether natural language or controlled vocabulary is used, the requirements of indexing a large body of literature are likely to mean that a large number of index terms will be required. Words are often quite long and are very variable in length. We have also seen, in talking about numeric data, the advantages of dealing with small fixed-length items such as numbers. Text systems therefore commonly employ file compression techniques in which the actual words are replaced by compact, often numeric, codes. This system is used in both the Medlars and UKCIS systems mentioned earlier [38,48].

To summarise, the problems of searching textual information are those of dealing with large files and a large vocabulary of variable-length items. Solutions to these problems involve file compression, careful file organisation, use of screening techniques, and so on. All these approaches are being elegantly combined by Heaps *et al.* [50].

The success of these techniques is testified by the growing success of computerised text search systems. The flexibility offered by computer searching, in particular such facilities as term truncation, syntactical linking, and nested Boolean logical linking, and its ability to cope with large volumes of data indicate it is likely to find increasing application in the future.

VI. Structural Information

As with the numeric data, the content of structural information in the secondary literature is limited, and we therefore have a situation in which many specialised systems have been developed to deal with structural data. Files of chemical structures, and means of accessing them, have been with us for a number of years, certainly from before computers. Traditionally, these have used notations, fragment codes, etc., as the means of recording structural information, and have allowed searches to be made for complete compounds or for all compounds containing certain specified substructures. More recently, topographical systems have been developed. In these the complete structures are recorded in the form of "connection tables", which store full details of all the atoms and bonds in a molecule, and the precise arrangement in which they are connected.

These topographical systems record the structural information more explicitly than the earlier methods, and so allow greater flexibility in substructure searching. They are, however, very expensive in terms of storage space, and they are specifically a product of the computer age since it is only with the advent of the computer that the ability to handle the necessarily large and complex files has become available. Progress in this field has recently been reviewed [51,52].

We have already seen (Table 2) that structural data is considerably more complex than either text or numeric data. Let us consider the problems in more detail.

Text is linear (*i. e.* one-dimensional) and unidirectional (*i. e.* words don't usually make sense backwards). Structural information on the other hand is three-dimensional (though the third dimension is not always significant of course) and multi-directional (*i. e.* a connected series of atoms makes chemical sense irrespective of the order in which they are considered). Text is made up from only one sort of component — characters, whereas structures are composed of two distinct types — atoms and bonds. The units of text (words) are also much smaller than those of structure (compounds). On a large file of chemical text the average word contained about 10 letters whereas in the Chemical Abstracts Service Compound Registry

System [53,54] the average compound contained a total of 80 components — 40 atoms and 40 bonds [55]. The range of components in text is fairly limited — 26 letters, 10 digits and a few special characters; whereas structures involve a wide range of components — over 100 atoms, various bond types, etc.

There are further substantial differences relating to the search facilities required. In searching text, there is generally only one starting point, namely the beginning of the word. (This is not true in systems which allow left-hand truncation, but even there the number of potential starting points will be limited). There will also be only one direction of pathtracing, and within the specified character string no alternatives will be required. In a structural record any point may be the starting point, path-tracing may proceed in any direction, including negotiating branch points and loops, and alternative atoms or bonds may be acceptable at various points in the required structures.

Obviously, then, it is very much easier to store and search linear conceptual information than multidimensional structural information. Consequently, computers are making a greater impact in handling structure files than for text. The emphasis is on the development of new systems and novel techniques, rather than on the computerisation of existing manual techniques. One such novel technique is that of set reduction [56].

We have already seen how the advantages of numbers from a computer-handling point of view have been applied to textual information by coding words as numbers. Similarly, the relative simplicity of linear information compared to structural has been one of the motivations behind the development of notation systems, which are nothing more than attempts to record structural information in a linear form (what I have called "pseudo-text"). It is worth noting that chemical nomenclature is therefore only a form of notation. To be slightly more rigid one should perhaps define a notation as a linear structural description algorithmically derivable from the complete structure. This definition would include systematic nomenclature, but not trivial names. A very comprehensive review of notation systems is available [57].

Briefly, notation systems attempt to record full, multidimensional structural descriptions in a linear form, by the use of more comprehensive symbols than atoms and bonds (*e.g.* symbols for particular chains, rings, functional groups). Thus, more information is recorded implicitly, in the *rules* of the notation, and less is recorded explicitly in the notations for individual compounds. The rules can therefore be quite complicated, in order to ensure the notations are unique and unambiguous. For the Wiswesser Line Notation, the rules are given in: Smith, E. G.: The Wiswesser Line-Formula Chemical Notation. New York: McGraw-Hill 1968. In this notation, for example, saturated carbon chains are simply indicated by an arabic numeral equal to the number of carbons in the chain, branching carbons by the symbol Y, carbonyl groups by V, and the end of a branch chain description by &. The notation for 5—ethylnonan—4—one:

$$CH_3 - CH_2 - CH_2 - CH_2 - \overset{\displaystyle CH_2 \atop |}{\underset{\displaystyle CH_3 \atop |}{CH}} - \overset{\displaystyle O \atop \|}{C} - CH_2 = CH_2 - CH_3$$

is thus 4Y2&V3, indicating the considerable compression which can be achieved.

Chemical structure files must be planned to cope with large numbers of compounds. The total number of unique compounds known has been estimated to be between 4 and 5 million, and the number of compounds in the CAS Registry System is now over 2 million. Great efforts must therefore be made to compress the files as much as possible. Thus, in the CAS and other topographical systems only non-hydrogen atoms and their connections are considered. The computer 'knows' the normal valence of every atom. Connections to non-hydrogen atoms are specified and all unfilled valencies are assumed to be to hydrogen.

This approximately halves the number of components per compound to be recorded. The information about hydrogen atoms is therefore *implicitly* recorded in the file. Greater compression still is achieved by using a linear description. Thus, for WLN it has been estimated [58] that the average notation contains 20—25 characters. However, this is only achieved by recording more information implicitly, since the full explicit description of a particular structure contains a finite number of bits of information and cannot be recorded in less than that number of bits. Unfortunately, there is an inverse relationship between the degree of implicitness and the flexibility of searching.

The searching of chemical structure files, especially topographical ones, is very tedious and expensive because of the number of alternative paths which have to be traced in looking for the required molecular fragment (see Table 1). Screening systems have therefore been particularly important in structure searching. In these, each structure on file has a screen associated with it in which are stored certain characteristics of the structure, for example, the total number of non-hydrogen atoms, the number of rings, etc. Many compounds can then be eliminated from a structure search merely by reference to the screen without the need to examine the structural record itself. For example, a search for the substructure:

cannot possible be satisfied by a structure which does not contain at least 11 nonhydrogen atoms, at least one ring, at least one chlorine, etc. Many different screening systems have been devised, and they have been applied to both topographical and notation files. They aim to screen out a large proportion of compounds so that the expensive full searching need only be done on a small portion of the file.

Summarising the contribution of computers to the handling of structural information one can say that they have made it possible to record full explicit structural descriptions of large numbers of compounds, thus allowing fully flexible substructure searches to be made. Equally, by allowing the development of screening systems they are making a significant contribution to the economics of such searches.

The KWIC indexing philosophy has also been successfully applied to chemical notations, in particular the Wiswesser Line Notation. The ease with which computers can handle numbers has also been exploited in handling structural data. In the various structural registries, notably the CAS Registry System [53,54], each compound is assigned a unique number which can then be used as the "label" for the compound rather than the full structural record.

At the moment these numbers are primarily used to link structural files with bibliographic or property files, but they are beginning to find wider usage. For example, they are now routinely included in *J. Org. Chem.* and if this development continues, the retrieval of structural information from the primary and secondary literature should become much easier.

VII. Summary and Conclusions

We have seen that there are essentially three different kinds of information in the chemical literature — (1) numerical, (2) conceptual, (3) structural. Computerised retrieval from the primary literature is as yet embryonic, and we must therefore turn our attention mainly to secondary sources. The normal secondary sources deal largely with words, *i. e.* conceptual information, the other types of information being dealt with by specialised systems. Computer search services therefore use machine-readable versions of the standard sources for current-awareness, specialised systems being used for retrospective searching.

The types of information present different problems for information retrieval, the order of difficulty increasing from (1) to (3). One way in which problems have been overcome is by translating information to simpler forms. Thus, textual information can be stored numerically, and structural information linearly or numerically. In addition, various techniques to improve file organization and matching techniques have been devised.

For the future, there is considerable speculation that more and more of the primary information, particularly numerical data, will be recorded in data banks, rather than in printed papers. Automatic data collection techniques will facilitate this. Thus, computer retrieval of primary information, though minimal at the moment, in the long term is likely to develop considerably.

However, even at this point in time, computerised retrieval from the literature is well established. Most advanced nations have at least one centre engaged in this type of activity. We are still in the early stages of development of computerised information retrieval but many exciting developments can be foreseen. Greater proportions of the primary and secondary literature will become available in machine-readable form. The treatment of numeric and structural data will continue to advance; the emphasis shifting more towards true information retrieval, and away from mere document retrieval. Thus, as the volume of literature continues (apparently unendingly) to grow, as the costs of computer processing continue to fall relative to human effort, and as the sophistication of the available computer techniques increases the machine will undoubtedly play an increasingly important role in information retrieval.

References

[1] Barnes, R. C. M.: Information use studies. Part 2 — Comparison of some recent surveys. J. Doc. *21* (3), 169—176 (1965).
[2] Menzel, H., *et al.*: Formal and informal satisfaction of the information requirements of chemists. U. S. Govt. Rept., PB 193,556 (1970).
[3] Cahn, R. S.: Survey of chemical publications. London: Chemical Society 1965.
[4] de Solla Price, D. J.: Little science, big science. New York: Univ. Columbia Press 1963.
[5] Baker, D. B.: Chemical literature expands. Chem. Eng. News *44* (23), 84—88 (1966).
[6] Baker, D. B.: World's chemical literature continues to expand. Chem. Eng. News *49* (28), 37—40 (1971).
[7] Barker, F. H., Kent, A. K., Veal, D. C.: Report of the evaluation of an experimental computer-based current awareness service for chemists. UKCIS Research Report No. 1, Chemical Society 1970.
[8] Callaghan, A., *et al.*: Students' chemical information project. Final report. OSTI (1969).
[9] Veal, D. C.: UKCIS, unpublished results.
[10] Survey of information needs of physicists and chemists. J. Doc. *21* (2), 83—112 (1965).
[11] Fugmann, R.: The theoretical foundation of the IDC-system: Six postulates for information retrieval. Aslib Proc. *24* (2), 123—138 (1972).
[12] Mooers, C. N.: Zato coding and developments in information retrieval. Aslib Proc. *8*, 3—21 (1956).
[13] Meyer, E.: Superimposed screens for the GREMAS system, E. Meyer, paper presented at the FID/IFIP Conference, Rome (1967). The IDC-System for Chemical Documentation. J. Chem. Doc. *9* (2), 109—113 (1969).

14) Guinness book of records. London: Guinness Superlatives Ltd. 1971.

15) Shannon, C. E.: A mathematical theory of communication. Bell System Tech. J. *27*, 379—423 and 623—656 (1948).

16) Lytle, F. E., Brazie, T. L.: Effects of data compression on computer searchable files. Anal. Chem. *42* (13), 1532—5 (1970).

17) Jurs, P. C.: Near optimum computer searching of information files using hash coding. Anal. Chem. *43* (3), 364—7 (1971). See also Computer searching of information files using hash coding. Comments, Lytle, F. E.: Anal. Chem. *43* (10).

18) UKCIS Search Manual. 1973 edition, UKCIS (in preparation).

19) Barker, F. H., Veal, D. C., Wyatt, B. K.: Retrieval experiments based on chemical abstracts condensates. Final report, UKCIS (1972).

20) An overview of worldwide chemical information facilities and resources. U. S. Govt. Rept., PB 176,160 (1967).

21) Veal, D. C.: Information systems for chemists. In: The applications of computer techniques in chemical research (ed. P. Hepple). Inst. Petrol. *1972*, 129—147.

22) Gushee, D. E.: Problems of the primary journal. J. Chem. Doc. *10* (1), 30—32 (1970).

23) Annual report to fellows, 1969—1970. Chemical Society 1971.

24) Kuney, J. H.: New development in primary journal publication. J. Chem. Doc. *10* (1), 42—46 (1970).

25) Mellon, M. G.: Chemical publications. Their nature and use. McGraw-Hill 1965.

26) The international compendium of numerical data projects, Codata. Berlin-Heidelberg-New York: Springer 1969.

27) Automated information handling in data centers, 2nd. edit. Report of the Codata task group on computer use, November (1971).

28) Scott, W. M., Ridley, R. G.: Computer techniques at the mass spectrometry data centre. In: The applications of computer techniques in chemical research (ed. P. Hepple). Inst. Petrol. *1972*, 148—154.

29) *e. g.* Computerised infrared retrieval search systems. Philadelphia, Pa.: Sadtler Research Laboratories Inc.

30) Cottrell, W. B., Buchanan, J. R.: Summary of environmental information activities of the nuclear safety information center. Report No. ORNL-TM-3009, Oak Ridge (1970).

31) Smith, E. G.: The Wiswesser Line-Formula Chemical Notation. McGraw-Hill 1968.

32) Codata Bulletin No. 4, Codata (1971).

33) Smith, F. J.: On-line data bank in atomic and molecular physics. Physics of electronic and atomic collisions. VII. ICPEAC. 1971. North Holland 1972.

34) Guide to procedures for the publication of thermodynamic data, IUPAC comm. on Thermodynam. and Thermochem. Pure Appl. Chem. *29* (1—3), 395—408 (1972).

35) Recommendations for the presentation of NMR data for publication in chemical journals, IUPAC Comm. on Mol. Struct. and Spectr. Pure Appl. Chem. *29* (4), 627—8 (1972).

36) Schwartz, E. S.: Methods of microglossary analysis. Data Acquisition and Processing in Biology and Medicine, Proc. 1964 Rochester Conf., 165—177 (1965).

37) Wyatt, B. K.: UKCIS, unpublished results.

38) Harley, A. J.: The medical literature analysis and retrieval system — Medlars. In: Computer-based information retrieval systems (ed. B. Houghton). London: Clive Bingley 1968.

39) Hutton, F. C.: RESPONSA, a computer search of a subject index, Proc. A. S. I. S. *5*, 121—4 (1968).

40) Lynch, M. F.: Univ. of Sheffield, private communication.

41) Barker, F. H., Veal, D. C., Wyatt, B. K.: Towards automatic profile construction. J. Doc. *28* (1), 44—55 (1972).
42) Batten, W. E.: UKCIS — The United Kingdom Chemical Information Service. Chem. Brit. *6* (10), 420—422 (1970).
43) Kent, A. K.: Performance and cost of "free-text" search systems. Inform. Stor. Ret. *6*, 73 (1970).
44) Key to Information, Universal Decimal Classification, FID Publ. 466, International Federation for Documentation, The Hague (1970).
45) Freeman, R. R., Atherton, P.: File organisation and search strategy using UDC in mechanised reference retrieval systems. Mechanised Information Storage, Retrieval and Dissemination, Proceedings of FID/IFIP Conference, Rome, June 1967. Amsterdam 1968, p. 122—152.
46) Salton, G.: Automatic information organisation and retrieval; Series in Computer Science. New York: 1968.
47) Sparck Jones, K.: Automatic keyword classification for information retrieval. London: Butterworths 1971.
48) Kent, A. K.: The UKCIS Inverted File Information Retrieval System, INFIRS. UKCIS Internal Report (1970).
49) Cawkell, A. E.: Citations in chemistry. Chem. Brit. *6* (10), 414 (1970).
50) Thiel, L. H., Heaps, H. S.: Program design for retrospective searches on large data bases. Inform. Stor. Ret. *8*, 1—20 (1972).
51) Chemical structure information handling, a review of the literature 1962—1968, National Academy of Sciences publication 1733. Washington, C. D. 1969.
52) Lynch, M. F., Harrison, J. M., Town, W. G., Ash, J. E.: Computer handling of chemical structure information. London: Macdonald 1972.
53) Leiter, D. P., Morgan, H. L., Stobaugh, R. E.: Installation and operation of a registry for chemical compounds. J. Chem. Doc. *5* (4), 238—242 (1965).
54) Morgan, H. L.: Generation of a unique machine description for chemical structures — A technique developed at chemical abstracts service. J. Chem. Doc. *5* (2), 107—113 (1965).
55) Leiter, D. P., Leighner, L. E.: A statistical analysis of the structure registry at chemical abstracts service; paper presented at 154th Nat. ACS Meeting. Chicago, Ill. 1967.
56) Sussenguth, E. H., Jr.: A graph-theoretic algorithm for matching chemical structures. J. Chem. Doc. *5* (1), 36—43 (1965).
57) Survey of chemical notation systems; National academy of sciences — National research council publication 1150. Washington, D. C. 1964.
58) Palmer, G.: Wiswesser line-formula notation. Chem. Brit. *6* (10), 442 (1970).

Received September 14, 1972

Identification of Organic Compounds
by Computer-Aided Interpretation of Spectra

Dr. Thomas Clerc

Laboratorium für Organische Chemie der Eidgenössischen Technischen Hochschule
Zürich, Switzerland

Dr. Fritz Erni

Naka Works, Hitachi Ltd., Ibaraki, Japan

Contents

I. Introduction

For structure elucidation and identification of organic compounds, physico-chemical methods have in many cases replaced the classical, purely chemical methods. Today's most important new methods include mass spectrometry, nuclear magnetic resonance spectroscopy (NMR), infrared spectroscopy (IR), and ultraviolet/visible spectroscopy (UV/Vis). The combined use of these methods has given most successful practical results, and a new working technique has been devised. Whereas formerely one used to employ a few carefully selected physico-chemical methods in investigating a given problem, interpreting the data in considerable depth, today almost every potentially useful method is employed but the data are interpreted rather superficially. This technique makes use of the fact that the various spectro-scopic methods give structurally significant results which are partly complementary and partly overlapping. The results thus provide mutual confirmation and may be interpreted without expert knowledge. This advantage of the new working technique is somewhat diminished by the much larger amount of data to be processed. Even in a medium-sized analytical laboratory the mass of data may be so vast that it is impossible to reduce and process it economically by conventional means. One therefore seeks to have a computer do as much of this work as possible. The basic problems connected with computerized data acquisition have been solved, at least in principle. Reduction of the mass of raw spectroscopic data to a few analytic-ally significant parameters still presents some unsolved problems, *e.g.* the definition of the analytically significant parameters and the represen-tation of their values in a form directly usable in a computer. In low-resolut-ion mass spectrometry this problem is comparatively easy to solve. The relevant parameters are the nominal mass and abundance of each fragment. These two values are easily represented as digital numbers. For the other spectroscopic methods considered here (IR, NMR, UV/VIS), the shape of a band can also be a useful, analytically significant parameter. However, its simple and concise representation by digital symbols is arbitrary and imprecise.

The problems which arise in the next step, the conversion of analytically significant parameters into structural information, and the attempted solutions are discussed below.

II. Interpretation of Spectra by Algorithmic Methods

Great difficulty is encountered in representing spectroscopical data in a concise form directly usable by a computer, and published algorithmic methods are restricted to the interpretation of mass spectra. Sophisticated aids to interpretation, as used with high-resolution mass spectra (*e.g.*

element maps, heteroatom plots, etc.) which are also produced by algorithmic methods, are not considered here.

A. Decision Vectors

The simplest method for the partial interpretation of mass spectra is to apply decision vectors.[20,21,22,25] Here the mass spectrum is perceived as a point in multidimensional space. Each mass number corresponds to one dimension. The abundance at this mass number (or a suitable function thereof) gives the component of the location vector along the corresponding coordinate axis. The more similar two mass spectra are, the less distance is there between the two points representing them in multidimensional space. If one infers that compounds of similar structure have similar mass spectra, then their representations in multidimensional space will cluster together. Therefore the mass spectra of a group of compounds which differ in one essential structural feature should, after transformation into points in multidimensional space, be separable by a suitable hypersurface in such a way that the points corresponding to mass spectra of compounds differing in this structural feature lie on opposite sides of the hypersurface. For practical reasons only hyperplanes are considered as dividing surfaces. Their equation is also interpreted as a vector. To determine on which side of the dividing hyperplane a given point lies, its location vector is scalar multiplied by the vector representation of the dividing plane, and the sign of the product gives the desired information. This simple mathematical operation, which can easily be carried out on a desk calculator, enables one to determine the group of compounds to which a given sample belongs. Such vector representations of dividing planes are therefore called decision vectors (cf. Fig. 1a).

The main problem obviously is to find suitable decision vectors. The most promising method is to employ the computer as a learning machine.[20] First, a set of mass spectra of known compounds is selected as the training set. Each spectrum in the training set contains information about the class to which the compound belongs. The learning machine is supplied with arbitrarily selected components of the decision vector and sequentially classifies every member of the training set. If the classification is correct, the machine proceeds to classify the next spectrum. If the classification is incorrect, the decision vector is changed so as to make it correct, using an equation along the lines of (1).

$$\boldsymbol{W'} = \boldsymbol{W} + c \cdot \boldsymbol{X} \tag{1}$$

$\boldsymbol{W}$: old decision vector
$\boldsymbol{W'}$: new decision vector
$\boldsymbol{X}$: incorrectly classified spectrum vector
c: correction factor

The correction factor c is given by

$$c = a \cdot \frac{\boldsymbol{W} \cdot \boldsymbol{X}}{\boldsymbol{X} \cdot \boldsymbol{X}} \tag{2}$$

where c, $\boldsymbol{W}$, and $\boldsymbol{X}$ are as above and a gives the magnitude of the correction step. If the spectrum vector $\boldsymbol{X}$ is to be classified correctly by the new decision vector $\boldsymbol{W'}$, a has to be greater than 1. With $a = 2$ the decision product of $\boldsymbol{X}$ with the new decision vector $\boldsymbol{W'}$ will be of equal magnitude but opposite sign.

This procedure is repeated until the decision vector is able to classify correctly all spectra in the training set. As the decision vector is corrected only when an error occurs, this iterative process slowly converges. A completely correct classification of the training set is possible only if the points are linearly separable (Fig. 1a). The outcome of this learning process, $i.e.$ the magnitude and sign of the components of the decision vector, depends very much on the starting values, the composition of the training set and the sequence in which the spectrum vectors are presented to the learning machine. This dependence can be used to detect components of the decision vector which are irrelevant for a given type of classification.[21]

A fundamental difficulty about this method of finding decision vectors is that the learning machine orients itself on the extreme points. The final position of the decision boundary is largely determined by spectra which are most atypical of their class. Therefore useful results can only be expected from a carefully selected, self-consistent and error-free training set.

Decision vectors have been published for determining the presence of heteroatoms, and the number of carbon and hydrogen atoms from low-resolution mass spectra. A sequence of appropriate vector-based decisions can determine the molecular formula of an unknown compound from its low-resolution mass spectrum alone.[20] However, the probability of obtaining the correct answer diminishes exponentially with the number of elementary decisions made. The present state of theory and practice in the application of learning machines and decision vectors to the classification of spectral data is such that the necessary reliability for longer decision chains exists only for very simple, exactly delimited classes of compounds.

If the two classes are not linearly separable yet do not overlap, correct classification may be achieved by what is called a "committee machine". Here a group of mutually independent decision vectors decides on the classification, the final result being given by a majority vote (Fig. 1b). If the two classes overlap, no reliable decision is possible (Fig. 1c).

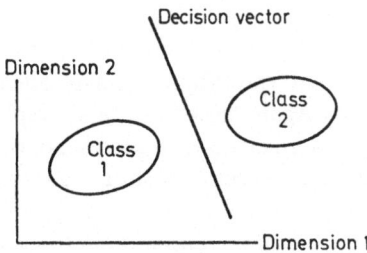

a) Linearly separable classes; simple classification possible

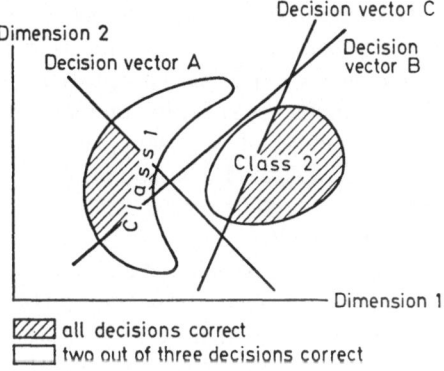

b) Not linearly separable classes correct classification possible by "commitee" machine

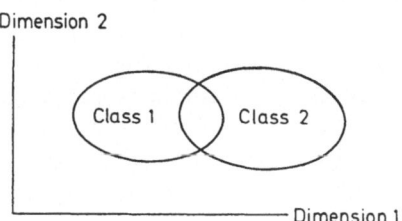

c) Overlapping classes, correct classification impossible

Fig. 1a–c. Geometrical interpretation of decision machines

B. Artificial Intelligence

The "artificial intelligence" method tries to simulate on a computer the decision process which takes place when a human analyst interprets a spectrum. [4–8,11,28,32,33,36,37,38)]

Based on a computer program which generates a complete nonredundant list of all possible structures for a given molecular composition; (Dendral) programs for the de-novo interpretation of low-resolution mass spectra have been developed.[7,8,11,33] Again however, their applicability is restricted to members of a very simple, exactly delimited set of compounds. At present, they can interpret only mass spectra of saturated, acyclic, monofunctional (SAM) compounds having not more than one heteroatom from the group containing O, N, and S, or simple aliphatic ketones. The first step of the program decides whether the given spectrum stems from a SAM compound; if it does not, the spectrum is rejected as not interpretable. If it does, the program identifies the heteroatom present and calculates the molecular formula; from this the Dendral algorithm generates all structures compatible with the given mass spectrum and ranks them according to plausibility. The best structures are then printed out. This program is extremely powerful when used within its limits. It is fairly complex even though it works only on a severely restricted set of very simple compounds. Its extension to even slightly more complicated compounds involves a tremendous increase in programming effort, so its present state probably is not far from the technical and economic limit.

Clearly, such methods are feasible only when the range is exactly delimited and not too broad and the unknown compounds are comparatively simple. This is the case with the sequence analysis of peptides, and several "artificial intelligence" programs, which give very useful results in practical work, have been developed for this application. [4-6,36-38] Here only a limited set of structural elements, the naturally occurring amino acids, have to be considered. They are always connected in the same way by an amide bond. Ramifications in the chain do not occur, and there are characteristic fragments for every possible structural element. The beginning and end of the chain are marked by easily recognized protective groups. The program first finds the terminal amino acid, as identified by the protective group, then combines this amino acid with all other amino acids and selects plausible combinations by searching for characteristic fragments exhibiting these particular combinations. This step is repeated until a complete self-consistent sequence is obtained.

III. Interpretation by Comparisons

Significant pointers to the structure of an unknown compound may be obtained by comparing its spectrum with a set of reference spectra and printing out the structure of the reference compounds whose spectra most resemble the unknown one. This method assumes the validity of the relation presented in Fig. 2. It avoids the main problems encountered with algorithmic methods of interpretation, namely that the field of application is

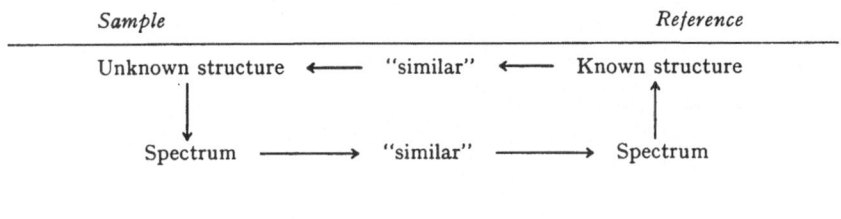

Fig. 2

limited and the complexity of the program increases exponentially with increasing depth of interpretation. However, there are other problems which restrict the performance of systems based on the comparison of spectra. Obviously, such a system is limited to the structures of the reference compounds included in the spectrum library. Such libraries have therefore to be rather extensive and their composition must reflect the user's sphere of interest rather closely. At the same time, core memory requirements and computer time increase at least linearly with the size of a data collection and may be quite considerable even for moderately sized libraries, so that some restriction of the spectral data is a "must". Another very important reason for using only a carefully selected subset of spectral, data is that the relation presented in Fig. 2 is not universally valid but applies only to certain spectral features, as indicated in Fig. 3. If we want a system for comparing

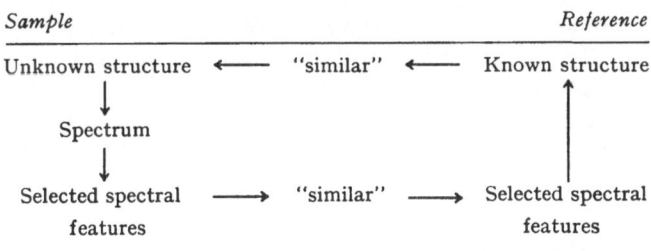

Fig. 3

spectra to give useful results, even when the unknown compound is not part of the library collection, by being able to retrieve analogous and/or homologous structures, the selected spectral features must conform to the more stringent relation given in Fig. 4. This also allows the system to exhibit

$$\text{Spectral feature} \left\{ \begin{array}{c} \text{not} \\ \text{slightly} \\ \text{strongly} \end{array} \right\} \text{ changing with } \left\{ \begin{array}{c} \text{identical} \\ \text{similar} \\ \text{different} \end{array} \right\} \text{ structures}$$

Fig. 4

a certain tolerance to the instrumental artefacts and human errors which can never be completely avoided in practical. The comparing systems described in the literature differ first of all in the spectral features selected. They differ further in the search strategy used and the method used to assign numerical values to different degrees of similarity between two spectra. Large computer-readable spectrum collections are available only in infrared spectroscopy and mass spectrometry, so that the existing systems center around these two analytical methods. Obviously, integrated systems using several different types of spectra simultaneously are highly desirable, but here the problem is how to handle incompletely documented compounds (unknowns and references). This important aspect is discussed below.

A. Systems for Comparing Mass Spectral Data

A simple method for quantitatively evaluating the similarity between two mass spectra consists in comparing the mass numbers of the most intensive peaks and counting the number of coincidences.[1,2,9,10,19,24,29,30,40,41] The higher this count, the more similar are the two mass spectra. But peak intensities tend to show rather low reproducibility and change in jumps although the structures differ only slightly; hence a system using this method is mainly suited for the retrieval of spectra of identical compounds. If the spectrum of the unknown compound does not form part of the library, one generally cannot expect useful results. Even if the unknown spectrum is included in the library, the system will normally not find the maximum number of coincidences because instrumental artefacts quite often exert a marked influence on the peak intensities, thereby changing the sequence, which is ranked according to intensity.

In the mass spectra of most organic compounds the peaks at low mass numbers contribute the largest proportion of the total ion current. As a result, this method discriminates against fragments which high mass numbers, which carry much of the structurally significant information. To overcome this difficulty, the spectrum may be divided into mass regions, from each of which the most intensive peak is selected. Its mass number then is used as a spectral feature. [18,19,29,39] The size of the mass regions m as well as the number of peaks p to be selected has been discussed.[24] The systems based on these features are designed mainly for retrieving identical spectra, so that the values for m and p may be chosen according to considerations based on information theory. Typical values include $m = 20$ amu and $p = 3$ peaks. If the mass region is taken to be 14 amu wide, the selection of the starting point becomes critical. Most mass spectra of organic compounds show clusters of peaks with centers of gravity within the range $m/e = 14n - 3$ to $m/e = 14n + 3$. It is advantageous to select a starting point which will not divide such clusters between two regions, [18,19] e.g. around $m/e = 14n + 7$.

Another way of trimming requirements is to omit the spectra intensity values completely and to retrain only the information "peak/no peak" at a given mass number. [16,17,42] To eliminate insignificant background peaks ("grass"), a threshold level may be specified. This is equivalent to coding intensity in only one bit. This method gives a computer-oriented code and permits very efficient programming. On the other hand, much relevant information is lost by this coding procedure. If the intensity is coded in more than one bit, the reliability of the system increases, but much more slowly than core memory requirements and computer time.

An interesting method for compressing mass spectral data is to sum the intensities of all peaks having the same modulo 14 mass number so as to give a shortened mass spectrum with only 14 peaks. [16] The intensity of the first peak corresponds to the sum of the intensities of the peaks at m/e 14, 28, 42, ... from the original mass spectrum; the second peak contains the intensities from the peaks at m/e 15, 29, 43, ... For most simple organic compound types these compressed mass spectra show very characteristic patterns. They are therefore very well suited to retrieve homologous compounds.

Another method which places most emphasis on the ability to retrieve homologous and analogous compounds is derived from an integrated system [14,15] which uses a set of spectral features optimized to the specification given in Fig. 4. These features are coded in one bit each. To keep the information loss within acceptable limits, individual threshold levels are used for every feature. The definitions used are presented in Table 1. Different

Table 1. Spectral features selected for the comparison of mass spectra

— Peak intensities at selected mass numbers, with 1 or 2 threshold levels
— Peak intensities in the modulo 14 reduced mass spectrum, [16] with 1 to 3 threshold levels
— Mass number at which the integrated total ion current rises above 99% of its maximal value, with 9 threshold levels
— Increase of the integrated total ion current in the lower, middle and upper third of the mass range, with 2 threshold levels respectively
— Contribution to the integrated total ion current from peaks at even mass numbers, with 4 threshold levels

weights are assigned to the four possible results of comparing two bits. The sum of the respective weights for the results of comparing all spectral features represents a very sensitive measure for the similarity between two mass spectra.

In all systems described in the literature the spectrum of the unknown compound has to be compared with every spectrum in the library. Even with libraries of moderate size, this uses up much costly computer time. To overcome this difficulty, one tries to preselect library spectra which are most unlike the spectrum of the unknown compound in order to skip them or break off the comparison prematurely. Hence the comparison process proper is preceded by a preselection algorithm, a filter, which the reference spectra have to pass before they are actually compared. The use of such filters can reduce computer time down to 10%.[19,24] To maintain the quality of the results, the filter must not reject any useful reference. Howewer, if only the spectrum of the identical compound has to be retrieved, this requirement need not be met very strictly.

Such a filter may consist of requiring the mass number of the most abundant ion in one spectrum to be present among the mass numbers of the most abundant ions in the other spectrum.[17,24] All reference spectra which do not satisfy this requirement are skipped. Likewise, similar molecular weight [19] or a similar modulo 14 compressed spectrum [19] may be specified as a prerequisite for the comparison. It is also possible to combine several different filters.[18,19] Another possibility consists in breaking off the comparison prematurely if intermediate results show marked dissimilarity. [14,15] This method requires the features most characteristic of the spectrum of the unknown to be compared first. If the features are individually weighted, the computer can determine the best sequence without human intervention. This method is equivalent to applying a dynamic filter which is automatically optimized to the spectrum of each unknown compound. It allows very strict preselection with a correspondingly high gain in program efficiency without impairing the quality of the results.

B. Systems for Comparing Infrared Spectra

All generally available collections of IR data in computer-readable form use the ASTM Coding which dates from the punched-card era. In consequence, most systems for comparing infrared spectral data are more or less stamped by this somewhat peculiar format, which uses band positions as the main spectral feature and assigns only limited weight to band intensities and shapes forms. Except in the fingerprint region (ca. 1500 cm^{-1} ... 1000 cm^{-1}), the feature "band position" adequately meets the requirement given in Fig. 4. Hence the published IR comparing systems [3,12,13,26,]

[27,35)] commonly produce good results even when the unknown compound does not form part of the library. They then give reference compounds which have identical functional groups and/or identical substructures.

If the ASTM Coding is transformed to a suitable one-bit code, extremely efficient comparing programs may be written, particularly if a machine-oriented language (Assembler) is used.[12)] However, this rules out the transfer of the program from one type of computer to another. The reference library may be grouped according to chemical criteria. A given group may, for example, contain only aromatic compounds, another only compounds having a carboxyl group. If the user knows to which group the unknown compound belongs, only the appropriate group has to be searched, giving a further gain in efficiency.[12)] The responsibility for selecting the correct group lies completely with the user: a wrong selection automatically excludes any useful result. Thus the user may unintentionally force the system to give results which conform to his expectations. Another method which saves large amounts of computer time uses the following scheme: the spectrum of an unknown compound is compared with a reference spectrum, and the latter is assigned a negative score for every mismatch. To allow for the variability of the spectral data, small deviations are not counted as a mismatch. The best reference spectra are obviously those which have received the least number of penalties. During the comparing operation a ranked list of reference compounds is continuously updated. Once a reference compound has accumulated more penalties than the currently worst reference compound in the list, the comparison with this compound is broken off, as this compound will certainly never be among the best references.

Inverted files have also been used for fast comparison of IR spectral data.[26)] Normal reference collections have an entry for every reference compound, containing information about the spectral features they exhibit. An inverted file has an entry for every spectral feature, containing a list of all compounds which exhibit this feature. Practically, every spectral feature has a byte string is assigned to it, in which every byte corresponds to a reference compound. If the reference compound exhibits the given feature, the value of the respective byte is set to 1. Otherwise its value is zero. To compare the spectrum of an unknown compound with the library, the byte strings corresponding to its spectral features are selected and added together. The result is an analogous byte string, in which the value of every byte gives the number of matches with that particular reference compound. Hence the bytes having the highest values indicate the best reference compounds. This search strategy allows for extremely, fast comparisons but also has some drawbacks and limitations. The core memory requirements are rather high, as one complete byte string has to be core-resident. If part of the byte string is moved to an external store, most of the time saving is cancelled out. The number of bits in a byte limits the number of features which can be compared in a single run, as an arithmetic overflow would give completely erroneous results. Therefore several spectral features are combined into one.[27] If criteria from statistical information theory are followed in selecting the features to be combined, the unavoidable loss in information content can be kept at an acceptable level. In a practical implementation the infrared spectrum is described by some 50 different features. However, as bytes having 3 bits each are used, not more than 7 of these features may be simultaneously used for comparison. Another basic drawback of inverted files is the difficulty of adding new reference compounds. In a conventional file, new reference compounds can be added by simply extending the file. With inverted files, additions necessitate rewriting the complete file.

C. Systems for Comparing other Types of Spectral Data (UV/Vis NMR)

As there are no computer-readable data collections of sufficient size commonly available, few significant papers have been published in this field.

A system for comparing UV/Vis spectra[31] uses correlation analysis. The correlation coefficient between two UV/Vis spectra interpreted as vectors is a very sensitive measure of similarity. By normalizing the vectors, the method may be made largely independent of concentrations. The limitations of this system are set rather by the high requirements as to the purity of the sample than by the algorithm.

Systems for comparing NMR spectra have been described only in connection with integrated systems [14,15], which combine several types of spectral data. A computer-readable collection of ^{13}C–NMR spectral data is being assembled.[23]

D. Integrated Systems for Comparing Combined Spectral Data

When data from several different spectroscopic methods are used for comparison purposes, greatly enhanced performance may be expected because the different methods complement each other. In principle, two or more specialized comparing systems may simply be combined. This however, creates two new basic problems. Firstly, it is extremely hard to devise efficient filters to exclude unsuitable reference compounds without discriminating against one type of spectroscopy. Secondly the unknown compound may not be fully documented with its spectral data. For example, the quantity of sample available may not be sufficient to run all spectra, or the recording of certain spectra may be impossible because of the physical and/or chemical properties of the sample. Every large reference data collection will contain incompletely documented compounds. On economic grounds one generally cannot insist that all compounds be fully documented, as sample preparation and spectra recording or retrieval of the relevant data from other collections are costly procedures. On the other hand, it is uneconomical to exclude otherwise useful reference compounds from the collection just because a few data are missing. An integrated system for comparing spectral data must therefore be able to handle incomplete data.

Besides spectroscopic features, other physical and/or chemical properties may be incorporated. A system has been described which uses data from mass spectrometry, IR, UV and ^{1}H—NMR spectroscopy as well as gas-chromatographic data, elementary composition and physical state under standard conditions.[34] A special comparing algorithm is used for every type of data, and each reference compound is scored for degree of match, with every data type. The best reference is the one which has the highest final score out of a possible 130 points. To this total the different data types contribute as follows: physical state 2 points; elementary composition 4 points; IR 36 points (two out of the three most intensive absorption bands at the same wavelength); ^{1}H—NMR 24 points (chemical shift of the most intensive signals equal); MS 40 points (mass number of three out of the four most abundant ions equal); gas chromatography 18 points (equal retention times on two standard columns; and UV 6 points (highest absorptions coincide). When no data are available for comparison, the reference compound is scored for an average match even though no actual comparison has been made. The reasoning behind this is that, while "no comparison possible" gives no hint as to the identity between the two compounds, neither does it speak against it. So incompletely documented compounds have a fair chance of being selected as a reference.

There has to be a special algorithm for comparing each data type and for evaluating the score, so that the program is not very efficient. The features selected hardly satisfy the requirement of Fig. 4. Hence the system is bestly suited for retrieving from a library a compound which is identical to the unknown. Experiments with a library of about 850 compounds have given results which can be judged useful when the limited size of the data base is taken into consideration.

Another integrated system which places heavy emphasis on retrieving analogous and/or homologous compounds will be described in rather more detail, as it contains some features not found in other systems.[14,15] It uses data from mass spectrometry, IR, UV and ^{1}H–NMR spectroscopy. From each spectrum features are selected which satisfy the requirement given in Fig. 4 as completely as possible. They are coded in one bit: feature present: code= 1; features absent: code= 0. For every reference compound the codes for all attributes are assembled into a bit string, which is called the signature of the respective compound. These signatures are organized into a matrix-like structure, as represented in Fig. 5. The signature of the unknown compound is generated by the same procedure. The trivial way to determine the similarity between two compounds is to count the number of corresponding bits having the same state. This number is entered into the column headed S (Fig. 5); at the end of the comparing process, this column is inspected. The reference compounds having the highest S values are the ones most similar to the unknown.

This rather crude similarity measure can be refined by weighting the spectral features according to their spectral and statistical significance

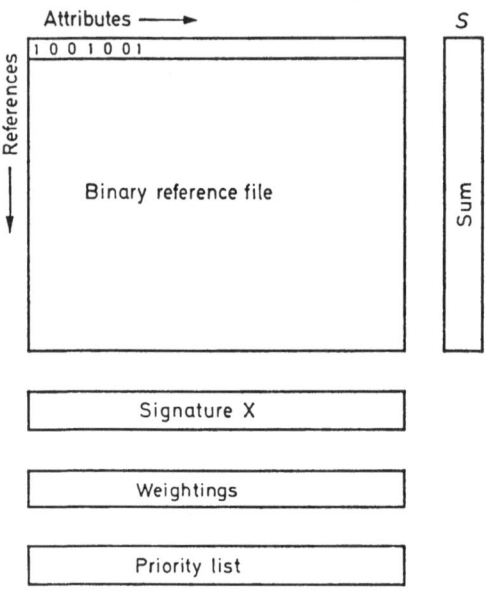

Fig. 5

(weight row in Fig. 5), and by differential evaluation of the four possible results of an elementary comparison. This results in a very sensitive quantitative measure of similarity. To improve efficiency, a filter is used to exclude extremely dissimilar signatures at an early stage of the comparing process Prior to the comparison process proper those signature elements which, in the case of identity, would contribute most to the similarity measure are determined by inspecting the signature of the unknown compound and the respective weightings. In this way a priority sequence may be determined for the features (priority row in Fig. 5). The elementary comparisons are now performed in the sequence given by the respective priorities so that the most characteristic features of a given unknown compound are automatically compared first. After a certain number of elementary comparisons, the actual value of S is tested against a threshold level. If it is below the threshold, the comparison is broken off. This method permits large savings in computer time without impairing the results. Determining the priorities according to the signature of the unknown ensures that the features most significant for the unknown compound are always used in the preselecting algorithm.

Incompletely documented compounds are dealt with in the following way: when data for the unknown compound are missing, the respective priorities are set to zero and no comparison is made. As the similarity measure S is printed out relative to what would be the value at identity, there are no problems in comparing S values. If some data concerning a reference compound are missing, instead of performing a comparison the computer scores it at a certain percentage (*e. g.* 85%) of the value for complete correspondence to ensure that premature breakoff does not occur as a result of comparisons not actually performed.

The features selected from mass spectra are given in Table 1. They contribute 120 bits to the signature. Proton magnetic resonance spectra are coded by defining partly overlapping ranges for the chemical shift with widths from 0.5 to 4 ppm to allow for the variability of the spectral data. Every range has three features, defined as follows:

(1) A signal is present in the respective range;

(2) a sharp singlet is present;

(3) a multiplet is present.

The results are no better when a more detailed description of the signal structure is given. The ^{1}H—NMR spectrum contributes 48 bits to the signature. In the IR spectrum ranges for absorption frequencies are defined, having a width of 50 cm^{-1} to 300 cm^{-1}. The most intense absorption band in every range is coded, its intensity being rated as low, medium, or high. If suitable values are selected for the weightings of the respective features, individual differences in assigning intensity codes will not have a marked effect on the results. The IR spectrum contributes 72 bits to the signature. The UV spectrum is coded over ranges 20 nm to 200 nm wide and assigned a rough intensity measure (low, medium, high).

Experiments with a library of about 1000 compounds have shown promising results, but the data base is still much too small to allow successful practical applications.

IV. Conclusion

Because of the limitations inherent in algorithmic methods of interpreting spectra, library search methods are likely to be used for the immediate future, except for special problems (*e.g.* Peptide sequence analysis). Nevertheless, the development of such algorithmic methods can be valuable in increasing our knowledge of the decision processes involved when a human analyst interprets a spectrum.

For successful application of library search systems, large computer-readable data files are a prerequiste. Such files should preferably contain the complete information from the original spectra. Today computers are being increasingly used for data acquisition in analytical laboratories, hence the production of such library files should soon become feasible as progress is made in defining significant spectral features and assigning balanced weights to them. Expert knowledge and much practical experience in spectroscopy is needed for this task, as well as familiarity with statistical information theory.

Systems where computer-aided interpretation of spectral data is used in order to identify the structure of organic compounds will find their main application in fields where samples have to be routinely analyzed without knowledge of their origin, *e. g.* in the fields of forensic chemistry, analysis of metabolites, and environmental pollution.

V. References

[1] Abrahamsson, S.: Science Tools *14*, 29 (1967).
[2] Abrahamsson, S., Ställberg-Stenhagen, S., Stenhagen, E.: Biochem. J. *92*, 2 (1964).
[3] Anderson, D. H., Covert, C. L.: Anal. Chem. *39*, 1288 (1967).
[4] Barber, M., Powers, P., Wallington, P., Wolstenholme, W. A.: Nature *212*, 784 (1966).
[5] Biemann, K., Kone, C., Webster, B. R.: J. Am. Chem. Soc. *88*, 2597 (1966).
[6] Biemann, K., Kone, C., Webster, B. R., Arsenault, G. P.: J. Am. Chem. Soc. *88*, 5598 (1966).
[7] Buchs, A., Delfino, A. B., Duffield, A. M., Djerassi, C., Buchanan, B. G., Feigenbaum, E. A., Lederberg, J.: Helv. Chim. Acta *53*, 1394 (1970).
[8] Buchs, A., Duffield, A. M., Schroll, G., Djerassi, C., Delfino, A. B., Buchanan, B. G., Sutherland, G. L., Feigenbaum, E. A., Lederberg, J.: J. Am. Chem. Soc. *92*, 6831 (1970).
[9] Crawford, L. R., Morrison, J. D.: Anal, Chem. *40*, 1464 (1968).
[10] Crawford, L. R., Morrison, J. D.: Anal. Chem. *40*, 1469 (1968).
[11] Duffield, A. M., Robertson, A. V., Djerassi, C., Buchanan, B. G., Sutherland, G. L., Feigenbaum, E. A., Lederberg, J.: J. Am. Chem. Soc. *91*, 2977 (1969).
[12] Erley, D. S.: Anal. Chem. *40*, 894 (1968).
[13] Erley, D. S.: Appl. Spectry. *25*, 200 (1971).
[14] Erni, F., Clerc, J. T.: Chimia (Aarau) *1970*, 388.
[15] Erni, F., Clerc, J. T.: Helv. Chim. Acta *55*, 489 (1972).
[16] Grotch, S. L.: Anal. Chem. *42*, 1214 (1970).
[17] Grotch, S. L.: Anal. Chem. *43*, 1362 (1971).
[18] Herlan, A.: Z. Anal. Chem. *253*, 1 (1971).
[19] Hertz, H. S., Hites, R. A., Biemann, K.: Anal. Chem. *43*, 681 (1971).

20) Jurs, P. C., Kowalski, B. R., Isenhower, T. L.: Anal. Chem. *41*, 21 (1969).
21) Jurs, P. C., Kowalski, B. R., Isenhower, T. L., Reilley, C. N.: Anal. Chem. *41*, 690 (1969).
22) Jurs, P. C., Kowalski, B. R., Isenhower, T. L., Reilley, C. N.: Anal. Chem. *41*, 1949 (1969).
23) Kellerhals, Hp.: Spectrospin AG, CH-8117 Fällanden, Switzerland, private communication.
24) Knock, B. A., Smith, I. C., Wright, D. F., Ridley, R. G.: Anal. Chem. *42*, 1516 (1970).
25) Kowalski, B. R., Jurs, P. C., Isenhower, T. L., Reilley, C. N.: Anal. Chem. *41*, 695 (1969).
26) Lytle, F. E.: Anal. Chem. *42*, 355 (1970).
27) Lytle, F. E., Brazie, T. L.: Anal. Chem. *42*, 1532 (1970).
28) Pettersson, B., Ryhage, R.: Anal. Chem. *39*, 790 (1967).
29) Pettersson, B., Ryhage, R.: Arkiv Kemi *26*, 293 (1967).
30) Raznikov, V. V., Talroze, V. L.: Dokl. Akad. Nauk SSSR *170*, 379 (1966).
31) Reid, J. C., Wong, E. C.: Appl. Spectry. *20*, 321 (1966).
32) Sasaki, S., Kudo, Y., Ochiai, S., Abe, H.: Mikrochim. Acta *1971*, 726.
33) Schroll, G., Duffield, A. M., Djerassi, C., Buchanan, B. G., Sutherland, G. L., Feigenbaum, E. A., Lederberg, J.: J. Am. Chem. Soc. *91*, 7440 (1969).
34) Schwartz, E. S.: J. Chem. Doc. *9*, 39 (1969).
35) Sebesta, R. W., Johnson, G. G., Jr.: Anal. Chem. *44*, 260 (1972).
36) Senn, M., McLafferty, F. W.: Biochem. Biophys. Res. Commun. *23*, 381 (1966).
37) Senn, M., Vonkataraghavan, R., McLafferty, F. W.: J. Am. Chem. Soc. *88*, 5593 (1966).
38) Sheldrik, B.: Quart. Rev. *24*, 454 (1970).
39) Stenhagen, E.: Chimia (Aarau) *1966*, 346.
40) Strogies, W., Tengicki, H., Vogel, L., Waldschmidt, H., Wittke, K.: Z. Anal. Chem. *245*, 76 (1969).
41) Talroze, V. L., Raznikov, V. V., Tantsyrev, G. D.: Dokl. Akad. Nauk SSSR *159*, 182 (1964).
42) Wangen, L. E., Woodword, W. S., Isenhower, T. L.: Anal. Chem. *43*, 1605 (1971).

Received December 6, 1972

Mass Spectra and Computers

Dr. Allan B. Delfino and Prof. Dr. Armand Buchs

Department of Physical Chemistry, University of Geneva, Switzerland

Contents

The mass spectrometrist sees the digital computer in quite a different light from the theoretical chemist. In mass spectrometry the computer is not generally used to carry out approximations by numerical techniques which vary the parameters. It is mainly cast in the role of mechanical controller, data acquirer and processer, file searcher and as resolver of structures from mass spectra of unknown compounds.

Thus, data acquisition and the use of computers to monitor mass spectrometers, in gas chromatography/mass spectrometry combinations for example, will not be covered in this chapter. Excellent review and original papers have been published in this field [1-12].

There are four general approaches to the computer-aided analysis of mass spectral data for structure determination. The first method, the use of file-searching techniques to correlate the mass spectrum of an unknown compound with reference spectra, really represents only a portion of the larger field of information storage and retrieval. Although file searching may be, at least for the times being, the most rewarding method available to the chemist for a rapid structure determination from spectral data, it will not be discussed here. For the status of this field of research the papers published by Hertz *et al.* [13] and by Erni and Clerc [14] make excellent reading.

The present chapter will be mainly devoted to the three other approaches for structure determination, *e. g.* the Learning Machine Approach (LMA), the Deduction Programming Approach (DPA) and the Heuristic Programming Approach (HPA). Much emphasis will be put on the artificial intelligence approach where the computer is used as a symbol manipulator. The last section will introduce a new departure in the use of a computer for mass spectrometry, the area of theory formation and the proposal of fragmentation mechanisms.

I. The Learning Machine Approach

The LMA work of Jurs *et al.* [15-23] in the field of mass spectrometry is important in applying the computer to structure identification. This work demonstrates what can be learned by the machine via a set of pattern dichotomizers or "binary pattern classifiers". The name binary pattern classifier comes from the fact that the pattern, in this case a mass spectrum, is placed in one of two categories. This name also implies that a classifier could, in fact, classify a pattern into one of many categories.

For our purposes, a binary pattern classifier is a computer program which is used to give a true or a false output when presented with a given mass spectrum. Thus the following question could be answered by either true or false:

Is oxygen present, in any functional form, in the substance which gave rise to this mass spectrum?

It was found that, by utilizing a sequence of such questions, empirical formulae could be determined from low-resolution mass spectra [15]. It took a series of 26 questions to determine an empirical formula of $C_7H_{16}O_3N_2$ or less. The method is not error-free, but the error can be as low as 5%.

The work done with the LMA approach in the area of infrared spectroscopy, melting points and boiling points in conjunction with mass spectrometry [18,19] will not be dealt with in any detail here.

A pattern classifier vector, V, is used to generate the true or false response by forming a dot product between itself and S, the mass spectrum vector.

$$c = V \cdot S$$

Experimental results showed that the best way to interpret the scalar result c was as shown below:

$$True : c \geqslant + 50$$

$$False : c \leqslant - 50$$

$$No\ answer : - 50 < c < + 50$$

A result in the 'No answer' range of c, the so-called 'dead zone' range, indicates the inability of V to dichotomize the pattern in terms of the question for which V was designed.

The components of V are derived by a process of 'machine learning' which is called training. In order to train the vector V, a set of randomly selected mass spectra, called the training set, is chosen. The number of components of the mass spectrum vector S has to be one larger than the greatest mass found in any of the spectra forming the training set. If, for example, 178 is the largest m/e value in the training set of spectra, all spectra will become vectors of dimension 179. The intensity of m/e i occupies the component S_i. The mass numbers are only implied by the existence of a component of S for each nominal mass. In our example, the position 179 is set to a value of 1 in all mass spectrum vectors of the training set, so that the 179th position of V will result in the position constant of the 178-dimension space function, where the decision plane, $i.\ e.$ the boundary hypersurface between the true and the false hypervolumes, will have the orientation described by the 178 first components of V. Various methods were studied to find the best way to transform the spectra for the training process. The best results were obtained when the spectra were modified in such a way that the intensities were transformed to the square root of the per cent of the base peak intensity. The dimensionality of the vectors was found to be reducible by various techniques such as graph theory [22], and double training for V [20].

The program used for training, the flow diagram for which is given by Jurs et al. [15], follows an error-correction procedure to arrive at acceptable values for the components of V. Initially all the V_i components have $+1$ (or -1) as their value. An initial dot product is then formed with the first mass spectrum in the training set. If the value of the dot product is outside the dead zone on the correct side, the response to the question, for example whether oxygen is present in the empirical formula, is deemed to be correct, and the next spectrum is processed. Whenever a spectrum is wrongly classified, a correction is made to V: a k multiple of the vector S is added to the old vector V in order to form a new vector V'.

$$V' = V + kS$$

The value of k is chosen so as to put the value of the dot product, $V' \cdot S$, on the correct side of the dead zone. Thus, this value is given by:

$$k = \frac{2}{|V|} (\pm M - c)$$

where M = size of the dead zone and c = original dot product of the incorrect answer.

New cycles through the complete training set are performed until no further corrections are necessary. The value of V will be found acceptable when all the spectra of the training set are classified correctly. After training is completed, predictive ability can be tested. Jurs [21] reports, for example, 98% predictive ability on a set of 300 mass spectra (89 from oxygen-containing compounds and 211 from substances containing no oxygen). The vector used on the 300 mass spectra had been trained on a different set of 300 mass spectra, 84 of them originating from oxygen-containing compounds. In order to achieve 100% recognition on the training set spectra, 375 feedback corrections were necessary. A similar process was unfortunately not carried out on the 26 vectors required to infer empirical formulae from low-resolution mass spectra. Such an experiment would, no doubt, be of great value in determining the potential of the LMA approach.

II. The Deduction Programming Approach

This section concerns the use of non-heuristic programs to deduce structural information from mass spectra. Automated systems were proposed as early as 1966 for compound-type identification [24,25] and as an aid in the interpretation of mass spectra [26]. Among them are the well-known polypeptide sequencing programs [27-29], the dialogue program designed by Biemann and Fennessey [30] and the subsequent man-computer interaction program for resolving complex chemical compounds [31]. The discussion will be restricted to the recent work we consider the most general and the most relevant.

Crawford and Morrison [32] have designed a general program to extract structural features from low-resolution mass spectra rather as a chemist does it. The program uses various interrogation schemes to yield successively the molecular mass, the presence of functional groups, and finally the molecular skeleton. Its success varies according to the complexity of the structure; a chemist might also have varying results if he was compelled to work only with low-resolution mass spectra as data. But the computer-controlled GC/MS systems, produce such enormous amounts of data that they must be processed by a computer. If it does not determine exact structures, at least the computer should reduce the possibilities, so that the mass spectrometrist can use this information to work out structures more quickly.

The main program deals with input and output requests and of various requests for structure determination. It coordinates the reading and the preprocessing of the mass spectrum, asks a subprogram for the molecular weight and the functional groups most likely to be present, calls on routines which specialize in various classes of mass spectra, calls for a nearest-neighbor table [33] if more specific information has not yielded reliable answers, and lastly asks *that output be made* of the results of all the other dispatches made previously.

The program for determining the most probable group was based on earlier work of Crawford and Morrison [34] in which average mass spectra, representative of various functional groups, were presented as points on a hypersphere. This program recognizes 12 classes of compounds:

> Aromatics,
> esters,
> ethers,
> acids,
> ketones,
> aldehydes,
> alkenes
> alkanes,
> alcohols,
> cycloalkanes,
> dienes and
> amines.

The molecular-weight routine assumes the highest m/e value to be near the molecular weight. If the molecular ion is not in the spectrum, the fact that it was not observed will be printed. The routine tries to distinguish between isotope peaks and the molecular-weight peak; if it decides that a certain peak is a likely candidate for the molecular ion, it checks whether there are any forbidden peaks in the mass range $M-3$ to $M-13$. A consistency check based on the greater stability of even-electron over odd-electron ions is also performed.

The empirical-formula routine uses the isotopic abundances and the inferred molecular mass and functional groups to arrive at upper limits for carbon content and degree of unsaturation.

If the inferred empirical formula is descriptive of a certain class of compounds whose fragmentation is known to the system, the special class routine for this type of compounds is brought in. If the class is not well known to the program, a general interrogation routine is enlisted for further processing.

The general interrogation routine does the 'inspired guesswork' of the system for classes which have no special class routine; it is based on the diagnostic key fragments proposed by McLafferty [35]. Associated with each key fragment is a piece of basic structure and some peak characteristics. Once the characteristics of the peak have been sensed, the basic structure is remembered for integration into the final result later on, this result is based on a probability of correctness calculated for each inferred basic structure. Finally, a structure-drawing routine prints the most probable structure.

To place their work in proper perspective, Crawford and Morrison compared the outcome of the program with the abilities of fourth-year students. The humans and the computer were about equal in their capacity to interpret the given mass spectra. The real significance of this work is that it is an *ab initio* approach, based not only on theory, but also on empirical evidence drawn from large numbers of reference mass spectra.

A method able to classify compounds into various functional categories from their low-resolution mass spectra, similar to the method used by Crawford and Morrison, has been recently proposed by Smith [36]. A large set of reference spectra is reduced to a correlation set of 'ion-series spectra'. Smith shows that each class of compounds that is available in an archive can be characterized by a typical ion-series spectrum. To determine the class of compounds to which an unknown substance should be assigned, the intensities of the signals found in its low-resolution mass spectrum are summed in the 14 different ion series calculated according to a formula similar to that used by Crawford and Morrison. Smith uses the following equation:

$$S_m = \frac{\sum_n I(30 + m + 14n)}{\sum I (j)}$$

where $I(j)$ is the intensity of mass j, $m = 1$ to 14, $n = 0, 1, 2, \ldots$ and where S_m represents the per cent contribution of ion series m to the total ion current.

This summation yields the ion-series spectrum for that compound, which is then matched, either manually or automatically, with the ion-series spectra of each class of compounds known to the system. Applications to environmental chemistry and to the analysis of complex geochemical samples seem to have been promising.

Although most of our discussion concerns programs that process low-resolution mass spectra, this does not indicate a bias on our part for low-resolution mass spectrometry. It is attributable rather to the fact that high-resolution work requires more in the way of instrumentation, so that low-resolution spectrometers are much more common. Hence people tend to focus their computing research toward the greater audience of low-resolution mass spectrometrists.

The work of Venkataraghavan, McLafferty and van Lears [37] is concerned with problems associated with structure determination on the basis of data acquired with high-resolution instruments. These authors present a generalized program to find structures from the elemental compositions of fragment ions. The program, as described in the publication mentioned above, could operate on ketones, esters, alcohols, amines and keto-alcohols containing saturated acyclic hydrocarbon moieties. The first part of the program tries to identify the general nature of the compound and its functional groups. Next specific subroutines search for known fragmentation pathways, typical of the compound class postulated. Much work is involved to identify a molecular ion. The candidate ion must contain the highest number of atoms of each element present in any ion of the spectrum, and its unsaturation degree must be zero or some positive integer. A consistency check is then performed by using the masses of the neutral fragments lost from the candidate molecular ion to yield the major fragment ions at the high-mass end of the spectrum. The number of occurrences of commonly lost fragments is weighed against those losses which are highly improbable. When this test rejects a postulated molecular ion, the program proposes candidates of higher molecular weight in a way similar to that proposed by Biemann and McMurray [38]. Losses of neutral fragments are further used to indicate the types of structures present, which are printed out by the computer.

The next stage of the program processes the information contained in characteristic ion series. Searches are conducted for homologous as well as for non-homologous important ion series. With that information, the system then attempts to deduce the most probable compound classes and then to postulate structures within the chosen classes. Using characteristic fragmentation pathways of a class, the program works its way back to a structure.

The power that can be built into routines for structure determinations restricted to particular classes of compounds is clear from the later discussion concerning the Inference Maker program. If the programs of the system proposed by Venkataraghavan *et al.* can be made as powerful as the Inference Maker, it could become the most effective of those reported to date.

III. The Heuristic Programming Approach

One of the most successful approaches to structure identification by means of computers was achieved at Stanford University with the *Heuristic Dendral Program*, based on the dendritic algorithm of Lederberg [39]. The Heuristic Dendral Program (HDP), as it was before 1970, is explained in various publications [40-42], and summarized in Fig. 1. The data, a mass

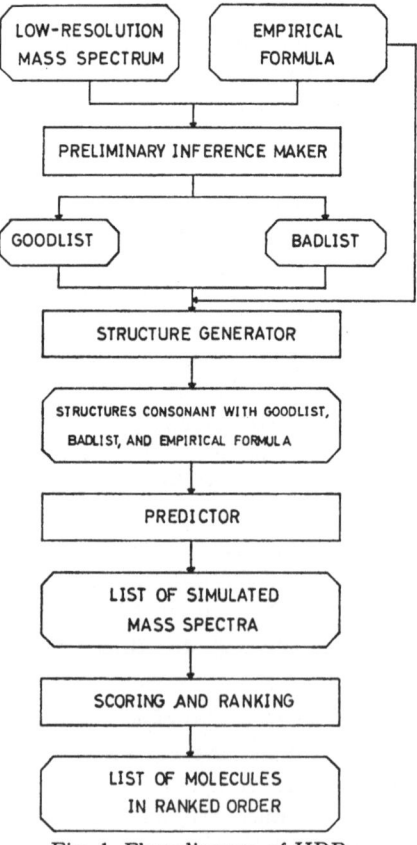

Fig. 1. Flow diagram of HDP

spectrum and an empirical formula, are given to the multi-phase HDP, and the planning phase, called the Preliminary Inference Maker (PIM), ascertains the presence or absence of the functional groups known to the program as a whole. The 'heuristics' used in this phase are similar to the rules that a mass spectrometrist applies to deduce the presence or absence of functional groups. These rules were given to the program in the form of tables. The results of the PIM phase are sorted into two parts: functional groups possibly present are put into a table called Goodlist and other functional groups known to be absent are put into Badlist. These tables are used as input to the next phase of HDP, called the Structure Generator (SG).

The SG uses the empirical formula along with Goodlist and Badlist to generate *exhaustively* but without redundancy a list of candidate structures for the form of the substance which gave rise to the mass spectrum. These candidates, consonant with the empirical formula, Goodlist and Badlist are then given as input to the third phase of HDP, the spectral simulator, called the Predictor.

After having generated characteristic mass spectral features from each candidate structure, the final phase of HDP, appropriately called the Scoring phase, compares the simulated mass spectrum and the given mass spectrum and scores the hypothetical structures according to the fit of the calculated mass spectrum to the actual mass spectrum. The candidate structures in order of descending score represent the final output of the HDP. If the program were correctly designed, the correct structure would have the highest score.

The HDP, operating as above, was tested with aliphatic ketones [41] and with aliphatic ethers [42]. This work underscored the lack of theory in the PIM phase. It was possible to generate unnecessarily large numbers of structures at great cost, only to have them ranked low or equal by the Predictor phase. As the size of the molecule increased, the SG phase was producing structures for the most part indistinguishable by their mass spectral behavior. This situation led to the evolution of the PIM program into the single-phase Inference Maker (IM) program which was tested on the mass spectra of saturated acyclic amines. Following the construction of the single-phase program with its repertoire of general subprogram units to determine amine structures [43], further and more powerful generalizations were made to give rise to the most recent program, summarized below, and described in one of the Dendral Project papers [44].

The current version of the Inference Maker program, which is written in the LISP programming language and also in FORTRAN, here in Geneva, can deduce the structure of any compound belonging to the class which is encompassed by the empirical formula $C_nH_{2n+v}X$ (X = O, S or

N, $v =$ valence of X). This class of Saturated Acyclic Monofunctional (SAM) derivatives includes amines, alcohols, thiols, ethers and thioethers. The emphasis was on keeping the program units as general as possible but tuning their 'heuristics' to ensure drastic pruning, yet never allowing the correct structure to be absent from the final output of candidates.

If the IM can deduce the nature of the heteroatom along with its molecular weight, the empirical formula can be calculated for this class of compounds, because they have a zero unsaturation degree. So, if the IM can filter out the mass spectra which do not originate from SAM compounds, the empirical formula can be precisely determined. Search methods for heteroatom and molecular weight are explained below. The rules governing the fragmentation modes of SAM molecular ions are general enough yet precise enough to enable the program to calculate the theoretical parameters required to elucidate structures of this class, thus relieving the chemist of this boring and time-consuming task.

The following discussion of how the IM processes mass spectral data is summarized in Fig. 2. At the outset, the filter program removes from the given mass spectrum all the ions belonging to the series of ions which could be used later on in the process of structure determination. These are the a-cleavage ion series, starting from all the masses of the general formula CH_2XH_{v-1}, $i.e.$ from 30, 31 and 47 where X is a nitrogen, an oxygen and a sulfur atom, respectively. The three alkyl ion series C_nH_{2n+1}, C_nH_{2n} and C_nH_{2n-1} are also removed, because of the role they play in the spectra of oxygen- and sulfur-containing SAM compounds. If this 'reduced' mass spectrum still contains any intense ions (10%), or if the remaining ions show, on the average, an abundance greater than 3%, the filter cancels further processing of the mass spectrum. A spectrum which is reduced enough to qualify as a possible SAM spectrum is then ready to undergo the hetero-atom-hypothesis process.

Using the a-cleavage ion series, the IM sums for each heteroatom the total intensity found in the corresponding a-series until the largest m/e value present in the spectrum is smaller than a given member of the homologous series. This intensity sum becomes the heteroatom's score. The heteroatom with the highest score is the one most likely to be contained in the empirical formula. The other heteroatoms are ranked in descending order of score and held for later processing, if required. If the first hereroatom does not give rise to a consistent structural output, the program might then be able to find its way out of a wrong choice of heteroatom due to some unforeseen weakness in the scoring method, or to some error in the mass spectrum. The rest of the program is more stringent in its requirements. Hence, an improper heteroatom should give rise to no output whatsoever. In fact, we had the program make bad first choices on purpose.

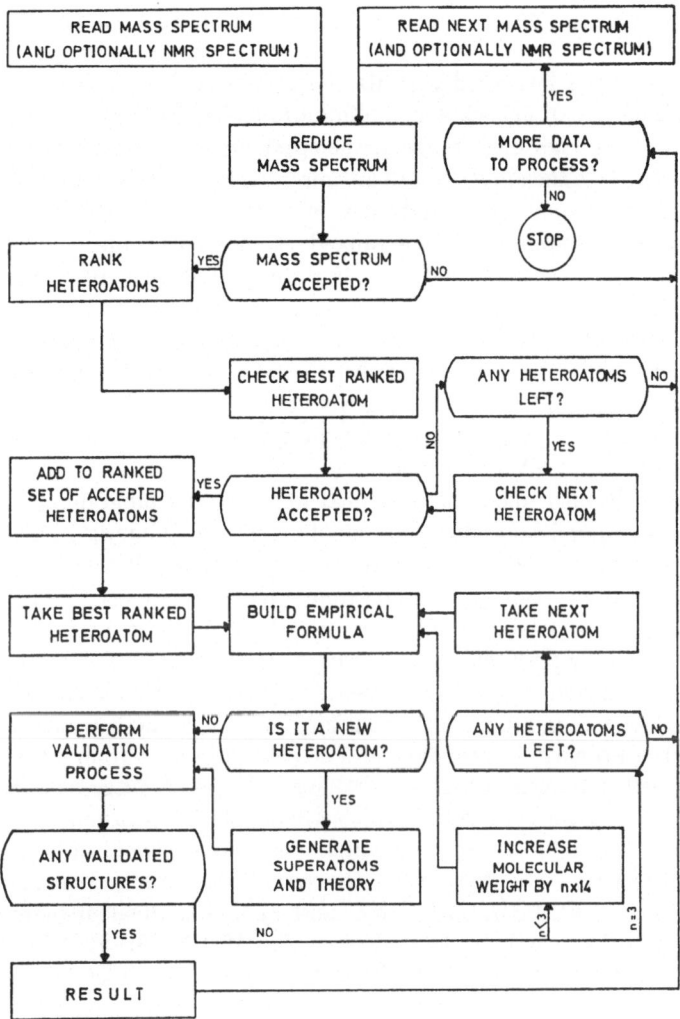

Fig. 2. Flow diagram of the IM program

It never found a structure corresponding to a false heteroatom and always went on with the process of validation until the correct heteroatom led to proper structure determination.

Intensities found in the alkyl ion series are also used to validate a heteroatom choice. Since nitrogen provides a better locus for the positive charge than either sulfur or oxygen, amine mass spectra carry a large proportion of the total ion current in the α-cleavage ion series, showing only a few

119

hydrocarbon-type ions. So a heteroatom will be kept for further processing only if the distribution of the ion current between the α-cleavage and the alkyl ion series is in accord with its charge retention capacity.

Once a heteroatom is chosen, a minimum molecular weight is inferred on the basis of the high mass range of the spectrum and the knowledge of the heteroatom. This then gives rise to an empirical formula. As with the heteroatom phase, the molecular weight determination phase is also a hypothesis stage, with the minimum value ranked as most plausible, and the next two higher homologs ranked after it. Failure to produce a structural output with one molecular weight causes the next higher homolog to be tried, until three values have been considered. Failure with all three values tried as molecular weights will cause a return to the heteroatom-hypothesis stage explained above. If no heteroatom, with any empirical formula, gives rise to an output, the program gives up and classifies the spectrum as not originating from a SAM molecule.

Having decided upon the empirical formula, the program then builds up a complete and nonredundant set of structural subunits called 'superatoms'. These superatoms contain the heteroatom and all the carbon atoms in the α positions. For each heteroatom, the complete set of superatoms represents every possible number of α-carbon atoms in accord with the valence of the heteroatom; and for each of these α-carbon atoms every type of substitution degree is represented. These superatoms are generated by the IM as combinations of the four latters M, P, S and T which stand for a methyl, a primary, a secondary and a tertiary carbon atom respectively. The defined canonical order of the symbols ($M < P < S < T$) demands that, in any sequence of symbols which will denote a superatom, any symbol must be to the right of a symbol which has a lower canonical value; thus, SPM is the only valid form from the set of the structurally equivalent forms: MPS, MSP, PSM, SMP, PMS, and SPM. For divalent heteroatoms only 12 groupings can be generated (P, S, T, PM, PP, SM, SP, SS, TM, TP, TS, TT), but for the trivalent nitrogen there are 31 different groupings which can be built. They are the twelve mentioned above, plus all the three symbol representing the tertiary amines: PMM, PPM, PPP, SMM, SPM, SPP, SSM, SSP, SSS, TMM, TPM, TPP, TSM, TSP, TSS, TTM, TTP, TTS and TTT. To specify the nature of the inferred heteroatom, a prefix is placed in front of the α-substitution symbol sequence. This completes the name of the superatom. These prefixes are AM, EA and TH for N, O and S respectively. Illustrated in Fig. 3 are some examples of superatom names along with the structures they represent.

The α-carbon atom environment, implied by the name of the superatoms, enables the IM to decide which mode of fragmentation would be exhibited by compounds with a particular superatom as the central structural subunit, after the alkyl radicals have been specified. Some of the rules

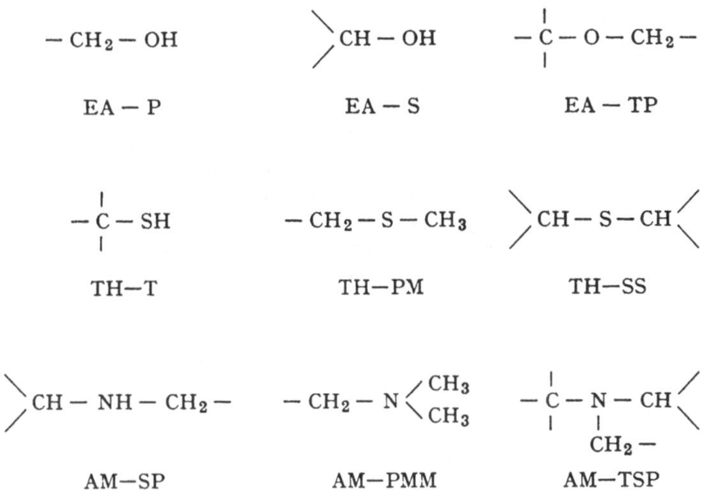

$$- CH_2 - OH \qquad\qquad {\Large\diagup\kern-1em\diagdown}CH - OH \qquad\qquad - \overset{\displaystyle |}{\underset{\displaystyle |}{C}} - O - CH_2 -$$

$$EA - P \qquad\qquad\qquad EA - S \qquad\qquad\qquad EA - TP$$

$$- \overset{\displaystyle |}{\underset{\displaystyle |}{C}} - SH \qquad\qquad - CH_2 - S - CH_3 \qquad {\diagup\kern-1em\diagdown}CH - S - CH{\diagup\kern-1em\diagdown}$$

$$TH-T \qquad\qquad\qquad TH-PM \qquad\qquad\qquad TH-SS$$

$${\diagup\kern-1em\diagdown}CH - NH - CH_2 - \qquad - CH_2 - N{\diagdown\kern-1em}_{CH_3}^{CH_3} \qquad - \overset{\displaystyle |}{\underset{\displaystyle |}{C}} - \underset{\underset{\displaystyle CH_2 -}{\displaystyle |}}{N} - CH{\diagup\kern-1em\diagdown}$$

$$AM-SP \qquad\qquad\qquad AM-PMM \qquad\qquad\qquad AM-TSP$$

Fig. 3. Structures of some superatoms and their corresponding names

which govern the fragmentation of a molecule are specific to the superatom contained in that molecule, others are generally applicable to all the super-atoms of a given heteroatom. The marked tendency of SAM molecular ions to fragment via α-cleavage is an example of a generally applicable rule.

The validation process begins with only those superatoms which can be contained in the empirical formula when a methyl group is attached to every free valence of the superatom. Superatoms requiring more carbon atoms than are available in the empirical formula are rejected out of hand. Early in the validation process an NMR spectrum can be optionally utilized to reject those superatoms which require more terminal methyl groups than are exhibited in the NMR spectrum. The number of heteroatom-bound methyl groups and the number of heteroatom-bound hydrogen atoms, deduced from the NMR spectrum, have also to be consistent with the structure of a superatom before it is retained for further processing.

There comes a time in the validation process when the surviving super-atoms will be transformed into general molecules by the specification of all possible sets of alkyl radicals, of unknown inner structure, which have to be attached to the free valences of the superatoms. To which of the free valences each possible alkyl radical may belong is not important at this point. The atoms of the superatom under consideration are subtracted from the empirical formula so far inferred, leaving the number of carbon atoms which must be distributed in all possible ways among the free va-lences of the superatom. For example, if the inferred elemental composi-

121

tion was $C_7H_{17}N$, superatom AM—SP would be extended to the four structures shown below (7):

$$H_3C \diagdown CH-NH-CH_2-C_3H_7$$
$$H_3C \diagup$$

$$H_3C-H_2C \diagdown CH-NH-CH_2-C_2H_5$$
$$H_3C \diagup$$

$$H_3C-CH_2-H_2C \diagdown CH-NH-CH_2-CH_3$$
$$H_3C \diagup$$

$$H_3C-H_2C \diagdown CH-NH-CH_2-CH_3$$
$$H_3C-H_2C \diagup$$

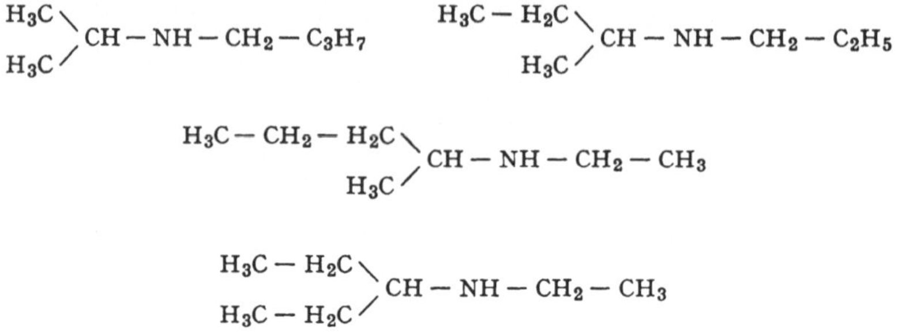

1

Each of these four general structures is then tested further, and the final output shows those which have passed all the tests related to a-cleavage, rearrangement ions, alkyl ions, etc. A complete description of the rules used to treat these molecules as unknowns can be found in one of our previous publications [44].

Diagram 1 shows the results printed by the IM when the *mass spectrum of 3-octanol* is processed without the help of an NMR spectrum. Once the program has inferred oxygen as the most plausible heteroatom, and $C_8H_{18}O$

Diagram 1. Output of the IM program with octane-3-ol as an unknown

---

ACTUAL MASS SPECTRUM = ((26 . 3) (27 . 31) (28 . 7) (29 . 35) (30 . 2) (31 . 32) (36 . 1) (38 . 1) (39 . 13) (40 . 2) (41 . 38) (42 . 7) (43 . 20) (44 . 9) (45 . 7) (51 . 1) (53 . 2) (54 . 1) (55 . 48) (56 . 1) (57 . 14) (58 . 9) (59 . 100) (60 . 4) (67 . 1) (68 . 1) (69 . 4) (70 . 3) (71 . 2) (73 . 1) (81 . 1) (82 . 1) (83 . 42) (84 . 4) (100 . 1) (101 . 16) (102 . 1) (112 . 2))

RESIDUE MASS SPECTRUM = ((51 . 1) (53 . 2) (54 . 1) (67 . 1) (68 . 1) (81 . 1) (82 . 1))

MASS SPECTRUM CORRECTED FOR 13-C = ((26 . 3) (27 . 31) (28 . 7) (29 . 35) (31 . 32) (36 . 1) (39 . 13) (40 . 2) (41 . 38) (42 . 6) (43 . 20) (44 . 8) (45 . 7) (51 . 1) (53 . 2) (54 . 1) (55 . 48) (57 . 14) (58 . 8) (59 . 100) (67 . 1) (68 . 1) (69 . 4) (70 . 3) (71 . 2) (73 . 1) (81 . 1) (83 . 42) (84 . 2) (100 . 1) (101 . 16) (112 . 2))

WAS A NMR SPECTRUM AVAILABLE	NO
INFERRED MOLECULAR WEIGHT	130
INFERRED EMPIRICAL FORMULA	C8H18O
GENERAL MOLECULES INFERRED:	
EA—S(C5H11,C2H5)	8 ISOMERS

---

as the empirical formula, the total search space contains 171 *a priori* valid structures, *i.e.* all the isomers having that empirical formula. Despite the fact that no molecular ion is present in the mass spectrum, the correct empirical formula is inferred by using the postulated minimum molecular weight value. If an NMR spectrum is used in conjunction with the mass spectrum, the presence of only two carbon-bound methyl groups and the absence of any oxygen-bound methyl radical allows the IM to reject straight away any superatoms with more than two free valences, and those superatoms with the symbol M in the name. This reduces the exhaustive, non-redundant set of twelve superatoms to only three. The three which survive the NMR tests are EA—P, EA—PP and EA—S, the last being the correct one. Moreover, the presence of only two methyl groups reduces the number of isomers for the general molecule EA—S—(C_2H_5, C_5H_{11}), shown in Diagram 1 and given below (2), to one of the eight isomers which are possible the structure of the C_5H_{11} radical is unknown.

$$\begin{array}{c} \text{OH} \\ | \\ C_2H_5 - CH - C_5H_{11} \end{array}$$

2

The IM program has been tested with 210 mass spectra, including amines, alcohols, ethers, thiols and thioethers. The correct structure was always included in the output and, even for large structures, the correct structure was often the only one proposed as an answer. Table 1 shows some results with various SAM compounds. Complete lists of the performance of the IM program are given in two previously mentioned publications [43,44]. The program is able to process a mass spectrum in about 3 seconds on an IBM 360, Model 65 computer. When an NMR spectrum is supplied along with the mass spectrum, the processing is even faster, because, as was explained above, superatoms can be rejected early on in the process, by means of simple tests. From the results so far attained, we feel that the IM program can determine SAM structures better than an experienced mass spectrometrist. The performance which was achieved should encourage the development of other programs for special classes of compounds.

Besides the heuristic programming approach oriented toward structure determination, there are two other challenging areas where heuristic programming is applied to mass spectrometry. Both these areas involve the use of the computer mainly as a symbol manipulator. Automated formation of a theory [45] and computer rationalization of mass spectral fragmentation mechanisms [46] are two topics of ongoing research that might eventually lead to a better formalization of the rules governing fragmentation under electron impact.

A. B. Delfino and A. Buchs

Table 1. Some IM program results

n	$C_nH_{2n+v}X$ compounds	Number possible $C_nH_{2n+v}X$ isomers	Number of inferred isomers A[a])	B[b])
	Amines			
6	Tri-ethyl	39	2	1
8	3-Octyl	211	26	2
9	Tri-n-propyl	507	2	1
10	Di-isopentyl	1.238	109	16
11	n-Undecyl	3.057	507	1
14	Di-n-heptyl	48.865	646	1
15	Tri-n-pentyl	124.906	40	1
17	N-methyl-bis-2-ethylhexyl	830.219	2.340	24
18	N-methyl-n-octyl-n-nonyl	2.156.010	15.978	1
20	N,N-dimethyl-n-octadecyl	14.715.813	1.284.792	1
	Ethers			
6	Ethyl sec-butyl	32	2	2
6	Di-isopropyl	32	1	1
7	Isopropyl sec-butyl	72	3	2
9	n-Butyl n-pentyl	405	8	1
20	Di-n-decyl	11.428.365	22.366	1
	Alcohols			
7	3-Ethyl-3-pentyl	72	1	1
8	2,3,4-trimethyl-3-pentyl	171	3	1
10	6-Ethyl-3-octyl	989	39	9
12	2-Butyl-1-octyl	6.045	1.238	25
14	3-Tetradecyl	38.322	1.238	1
	Thioethers			
7	Isopropyl sec-butyl	72	4	3
8	Ethyl n-hexyl	171	8	1
8	Di-sec-butyl	171	3	1
10	Di-n-pentyl	989	12	1
14	Di-n-heptyl	38.322	153	1
	Thiols			
6	1-Hexyl	32	8	1
6	2-Hexyl	32	12	5
8	1-Octyl	171	39	1
10	1-Decyl	989	211	1
12	1-Dodecyl	6.045	1.238	1

[a]) A = Inferred isomers when only mass spectrometry is used.

[b]) B = Inferred isomers when the number of methyl radicals is known from NMR data.

A. The Meta-Dendral Project (MDP)

The MDP being undertaken at Stanford University seeks to write a program which, given structures and corresponding mass spectra, will be able to discover the parts of a unifying theory of mass spectrometry. The program starts with a minimum of assumptions. When enough of the parts have been identified, it will unify them into a description of how molecular ions fragment under electron impact. The overall strategy of the MDP program can be divided into three parts:

1. Given a set of operations called, for example, Breakbond, Migrategroup, etc., etc., apply them in all possible singlets, doublets, triplets, up to some optional multiplet, to the given input structure. The data are the low-resolution mass spectrum and the corresponding structure. The output will consist of ionic structures and the sequence of processes that could give rise to them. These outputs will become inputs to the second part of the program. This step in no way uses an existing theory, which would be a form of question-begging for a program such as this. No information exists about charge loci, radical site triggering, or new bond formations as tools for this heuristic search.

2. Given the output of stage 1 for many related compounds, the program will generalize the results to find which sequences of processes exist in all the related compounds. Then it will search the retained ionic structures, for the common factor that allowed the process to occur. This distilled essence, along with the process set, becomes input to the final stage of the MDP program.

3. The output of stage 2 must finally be integrated into the existing theory. At the outset, this will, of course, be zero. But as soon as two structural essences and two process sets exist, it must be determined whether the second is just a corollary of the first, or vice versa, or whether they are disjoint situations with their associated actions. The theory will then take the form of these so-called situation-action rules [45]. The situations are the structural essences and the actions are the process sets.

Since the project has only just started, no definite chemical results have yet been published. But the explanation of the program, in a paper addressed to computer scientists [45], gives hope that the project will soon bear some encouraging results for chemistry.

B. The Ion Generator Program (IG)

The approach taken at the University of Geneva [46] addresses quite a different problem. We have devised a heuristic search program, the *Ion Generator* program, to simulate the formation of ions in a mass spectrometer.

It is based on the well-known method of electron bookkeeping, widely used by mass spectrometrists to explain the formation of a particular ion from its precursor ions. The IG generates, exhaustively, all motivated sequences of electron book-keeping processes, and heuristically selects those sequences, called mechanisms, which might help to formalize mass spectral fragmentations. It is in its exhaustiveness that the program differs from the chemist. The chemist is prone to stop after he finds one reasonable mechanism; the the program will find all of the reasonable mechanisms.

The input to the IG program is a structure and its mass spectrum. The output is a list of ions with their masses and structures, and with the mechanisms which gave rise to each ion generated. The output also includes, for each ion, the intensity with which it appears in the actual mass spectrum. We call these electron book-keeping processes primitive operations; a sequence of primitive operations constitutes a mechanism. For some of the generated ions, only one mechanism will be proposed; for other ions, many sound mechanisms may result. Some mechanisms will lead to the formation of ionic species which are absent from the mass spectrum. A mechanism which leads to an ion that was not present in the mass spectrum, and whose daughter ions are also missing, could be a candidate for rejection. Of course, the real test for mechanism rejection would be experiments with labeled compounds.

In generating molecular ions from molecules, and fragment ions from molecular ions, the program applies only the primitive operations which are *motivated* by certain structural environments that exist in the subject structure. It does not perform every primitive operation in every situation. The IG program rests on the assumption that the given motivations to perform the primitive operations are a sound basis on which to work. In its current implementation, the IG will perform the following primitive operations, given the proper structural situation:

> Ionization of the structure with charge localization.
> Homolysis of bonds in the β positions of an existing radical site.
> Bond formation by the pairing-up of two adjacent radical sites.
> Transfer of an atom or of a group of atoms to an existing radical site via cyclic transition states of various sizes.
> Ring closures between existing radical sites or between one existing radical site and another radical site created by a simultaneous homolysis.

It is envisaged that these primitive operations will be only a part of the complete set of operations used by the program in the future.

In order to generate sequences of primitive operations in an exhaustive and nonredundant way, the IG program first checks that the required structural motivations exist for all primitive operations in the molecular ion. These motivated operations result in transformed structures. Some of these resultant structures will be fragment ions, but others will be rearranged forms having the same mass. The rearranged ions will undergo the same check for structural motivations for all the primitive operations. Hence, all possible sequences will be applied to the molecular ion to result in every possible daughter ion. The nonredundance results from the fact that only one motivated operation is done at one time on any of the intermediate structures.

At present, the program is restricted to generating only primary ions. However, nothing in the generation scheme prevents the program from going further; it is so restrained because of the amount of output that is generated. When more is learned about methods which can justifiably be applied to limit the amount of output, secondary and higher order ions will be generated also.

The motivations and the restrictions put on the primitive operations are summarized in Diagram 2, and will briefly be described below.

Diagram 2. Application rules of the primitive operations of the IG program

Primitive operations	Motivations	Restrictions
Ionization		Selected ionization threshold range
Homolysis	Bond β to a radical site: formation of a new bond	Bond was formed in a previous step of the ongoing mechanism
Makebond	Presence of two adjacent radical sites	Bond was broken in a previous step of the ongoing mechanism
Transfer of R. R = H or radical	Presence of a radical site. Appropriate size for a cyclic transition state	No blatant strain in the transition state. Shifted bond must not have been formed in a previous step of the ongoing mechanism. Selected maximum number of successive transition states must not be exceeded
Ring closure	Two existing radical sites at proper distance. Intramolecular attack of a bond due to reactivity of the radical site. Formation of a new bond	Original broken bond must not have been formed in a previous step of the ongoing mechanism. Allowed sizes for the rings formed. No blatant strain in resulting ring

It is well known that the most successful explanation of the fragmentation which occurs under electron impact still relies on the concept of charge localization, just prior to fragmentation. This concept is accepted by most mass spectrometrists, and its importance is underlined in textbooks dealing with organic mass spectrometry. Our program has been designed to recognize that the site on which the positive charge is best accomodated is structure-dependent. Functional groups containing heteroatoms will be preferred as charge loci over carbon-carbon sigma bonds or carbon-carbon double bonds. For each different type of electrons (n, π, σ), and for each environment, an ionization threshold value has been semi-empirically chosen. Nitrogen is known to accomodate the positive charge better than sulfur or oxygen. So, in the sequence of ionization threshold values, ionization on a nitrogen atom will have priority over ionization on a sulfur atom or on an oxygen atom. Table 2 shows ionization threshold values selected for a number of structural environments. The user of the program is free to choose the complete range of ionization threshold values, or to

Table 2. Ionization threshold values of the IG program

Structural environment	Ionization threshold value
$>$N—	1000
$>$NH	950
—NH$_2$	900
$>$C=S	850
$>$C=O	800
—S—	750
—O—	700
—SH	650
—OH	600
—C≡C—	550
$>$C=C$<$	500
$>$C$<$	450
$\geq$CH	400
$>$CH$_2$	350
—CH$_3$	300

select a smaller range with optional upper and lower limits. If, in a structure containing a keto group, one wants ionization to occur only by removal of a lone-pair electron from the carbonylic oxygen atom, the ionization threshold value range can be chosen from 750 to 850. On the other hand, difunctional molecules can be ionized either on both heteroatoms in the same run or on each heteroatom separately in two successive runs. This affords an interesting possibility to study the additivity property of mass spectral fragmentations. When a rather broad range of ionization threshold values is allowed, the program builds and fragments the molecular ions in the order of the priority scale shown in Table 2.

The only homolytic cleavages which are presently allowed to occur are those involving bonds which are in the β positions with respect to an existing radical site. The motivation to make a fission is the potential to form a new bond by radical pairing, either in the ionic or in the neutral species. This is known to be an energetically favorable action. Around a radical site, every bond which is in a β position is fissionable, unless it has been formed by the IG program in a previous operation of the ongoing process. Homolysis of β bonds will also be suppressed if the resulting intermediate or final species contains a number of radical sites exceeding the number which has been chosen by the user. The formation of a bond by radical pairing will occur if the resulting bond is not one which has been broken by the IG in a previous step.

Another way to form a new bond at a radical site is by intramolecular attack of a bond via a cyclic transition state. This leads to transfer of a hydrogen atom or of a group of atoms to the radical site, thus creating a new radical site able to trigger either homolysis or a new cyclic transition state. The user chooses the number of transition states allowed to occur in any sequence of primitive operations, as well as the sizes of the rings. Most of the time we allow only two or less successive cyclic transition states to occur in a mechanism. The available sizes of the transition states range from four-membered to eight-membered rings. The Ion Generator will not attempt to go through blatantly strained transition states. Another interesting feature of the IG is that it distinguishes between 'long-lived' and 'short-lived' transition states. When the first transition state applied is long-lived, the spatial conformation of the structure is retained until the second transition state has been completed. This allows two transition states to occur simultaneously, in a concerted way.

The program knows that intramolecular attack of a bond by a radical site does not always lead to transfer of an H atom or of a radical, but can be followed by cyclization with simultaneous elimination of the radical which was attached to the attacked bond; ring structures are thus obtained. An example of such a ring formation is given in Scheme 2 (see p. 134).

To illustrate the performance of the IG, some examples of mechanisms proposed for the formation of primary ions from compound 3 will be given.

$$
\begin{array}{c}
\qquad\qquad\qquad\qquad\overset{\displaystyle O}{\overset{\displaystyle \|}{}} \\
CH_3-CH_2-CH_2-CH_2-CH_2-CH_2-CH_2-CH_2-\underset{\displaystyle |}{CH}-C-O-CH_3 \\
\qquad\qquad\qquad\qquad\qquad\qquad CH_2-CH_2-CH_2-CH_2-CH_2-CH_3
\end{array}
$$

3

The data submitted to the IG were the structure of *3* and its low-resolution mass spectrum, recorded at 70 eV. The input of structures and mass spectra is explained in our latest publication [46]. The way structure *3* was numbered for input is shown below (*4*).

$$
\begin{array}{ccccccc}
18 & 17 & 16 & 15 & 14 & 13 & 19 \\
C-\!\!-C-\!\!-C-\!\!-C-\!\!-C-\!\!-C & & & & & & O
\end{array}
$$

$$
\begin{array}{cccccccccccc}
& & & & & & & & & & & \| \\
C-\!\!-C-\!\!-C-\!\!-C-\!\!-C-\!\!-C-\!\!-C-\!\!-C-\!\!-C-\!\!-C-\!\!-O-\!\!-C \\
1 & 2 & 3 & 4 & 5 & 6 & 7 & 8 & 9 & 10 & 11 & 12
\end{array}
$$

4

The numbering scheme can be arbitrary; different numbering schemes for the same structure will require different input cards, but will in no way affect the resultant structures and mechanisms proposed by the IG. Currently the program can accept up to fifty atoms, not including the hydrogen atoms.

The IG returns two kinds of output. On the main output, every step of the processes involved in the formation of an ionic species is shown in detail, with the structures of both the ionic and neutral (if any) species printed in linear notation. The linear notational scheme used for output is a non-canonical form of the Dendral notation of Lederberg [39] with some minor modifications that we have mentioned in our latest paper [46].

The ionization range selected to ionize structure *3* extended from 800 to 1000. Thus, ionization took place only by removal of one of the lone-pair electrons from the carbonylic oxygen atom. A short-hand notation of the output for that process is:

ION 19   IMASS 270   INT 16%

Which means that ionization affected atom no. 19, yielding an ionic species with *m/e* 270, appearing in the actual mass spectrum with a relative abundance of 16%.

From that molecular ion the program generated altogether 24 ions with different *m/e* values. Table 3 shows a list of the ions generated along with the number of mechanisms proposed for each ion. For demonstration purposes, the sizes of the allowed transition states were restricted to 4, 6 and 8-membered rings, and the ring closures to 4-membered rings. No restrictions were placed on the size of the radicals allowed to be transferred. A few examples will suffice to illustrate some of the proposed fragmentation mechanisms.

Table 3. Primary ions proposed by the IG program for 2-*n*-hexyl-
methyl-*n*-decanoate

m/e	Number of mechanisms proposed	Relative abundance in mass spectrum (%)[a]
270	1	16,0
255	8	0,4
242	31	0,6
241	2	2,5
239	1	4,8
227	6	2,7
214	19	1,1
213	2	5,9
199	8	15,0
186	19	68,0
185	2	7,7
171	9	19,0
158	16	100,0
157	2	14,0
143	5	30,0
130	6	2,5
129	1	17,0
115	10	28,0
102	2	2,6
101	2	22,0
87	7	99,0
73	1	2,9
59	1	7,9
45	4	2,9

[a]) Mass spectrum no. 1898 from MSDC catalogue.

1. Simple α-Cleavages Yielding *m/e* 239 and *m/e* 59

The IG rationalizes the formation of these two ions in the following way:

ION 19   HOM 10,11   MKB 10,19   IMASS 239   NMASS 31   INT 4,8%

This reads: "Ionization on atom no. 19, homolysis between atom no. 10 and atom no. 11, bond formation between atom no. 10 and atom no. 19". The indications about the masses of the ionic (IMASS) and the neutral (NMASS) species, as well as about the relative abundance retrieved from the mass spectrum are presented in addition to the mechanism.

For *m/e* 59 the sequence of primitive operations forming the α-cleavage mechanism is presented as follows:

ION 19   HOM 9,10   MKB 10,19   IMASS 59   NMASS 211   INT 7,9%

The structures of the ionic species and of the neutral species also appear in the output. The structures for the second mechanism described above are presented as shown below, indicating the site of the positive charge and of the radical.

Ionic species

$$O(+)\equiv C-O-CH3$$

Neutral species

C(.)H——CH2—CH2—CH2—CH2—CH2CH3CH2—CH2—CH2—CH2—
CH2—CH2—CH2—CH3

2. McLafferty Rearrangement Ions

Among the 19 mechanisms proposed for the formation of m/e 186 and among the 16 mechanisms generated for m/e 158, the two mechanisms correspond- ing to McLafferty rearrangements are present. For the rearrangement involving the *n*-hexyl group, the one which yields m/e 186, the program did the following:

ION 19   6SMRTH-14   HOM 9,13   MKB 13,14
IMASS 186   NMASS 84   INT 68%

The notation '6SMRTH-14' translates to: 'Short-lived 6-*M*embered *R*ing *T*ransfer of a *H*ydrogen atom abstracted from atom no. *14* (to the existing radical site)'. The corresponding structures of ionic and neutral species are:

Ionic species

O(+)H=C——C(.)H—CH2—CH2—CH2—CH2—CH2—CH2—CH2—CH3O—CH3

Neutral species

CH2=CH—CH2—CH2—CH2—CH3

When the McLafferty rearrangement affects the *n*-octyl group, the ion with m/e 158 is formed. The mechanism proposed by the ION GENERATOR is:

ION 19   6SMRTH-7   HOM 8,9   MKB 7,8
IMASS 158   NMASS 112   INT 100%

It is well known (47) that in the fragmentation of normal long-chain methyl esters the formation of fragments of the general formula $(CH_2)_n COOCH_3^+$, where $n = 2, 6, 10 \ldots$, is favored. Such ions are indicated at m/e 87, 143, 199, 255 etc. in the mass spectra of unbranched methyl esters, *e. g.* spaced at intervals of 56 mass units. To explain their genesis, Djerassi *et al.* [48] propose mechanisms which involve two successive transfers of hydrogen atoms. With our example (*3*), among the mechanisms generated for m/e 171 and for m/e 199, one finds the two mechanisms which are analogous to the one proposed for the formation of m/e 87 in the mass spectra of unbranched long-chain methyl esters.

m/e 171

ION 19   8SMRTH-5   6SMRTH-9   HOM 7,8
IMASS 171   NMASS 99   INT 19%

This corresponds to the mechanism depicted in Scheme 1. Exactly the same

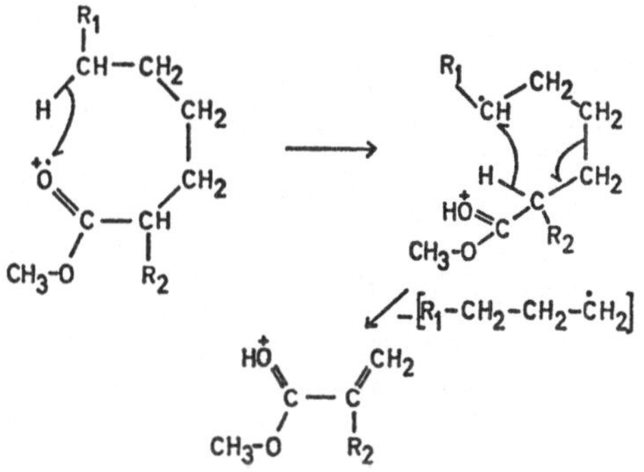

$R_2 = H \longrightarrow m/e\ 87$
$R_2 = C_6H_{13},\ R_1 = C_4H_9 \longrightarrow m/e\ 171$
$R_2 = C_8H_{17},\ R_1 = C_2H_5 \longrightarrow m/e\ 199$

Scheme 1

mechanism is proposed for m/e 199, with the difference that the transition state takes place in the *n*-hexyl radical. The mechanism is:

ION 19   8SMRTH-16   6SMRTH-9   HOM 13,14
IMASS 199   NMASS 71   INT 15%

As an example of a mechanism proceeding through a 'long-lived' transition state, one can consider the expulsion of a *n*-propyl radical from long-chain methyl ester molecular ions. Djerassi *et al.* [49] have postulated a mechanism which accounts for the observation that the three expelled carbon atoms are the closest neighbors of the carbonylic carbon atom, *i. e.* the carbon atoms numbered from 1 to 3 in the structure shown below.

$$H_3C-(CH_2)_n-\underset{3}{CH_2}-\!\!-\underset{2}{CH_2}-\!\!-\underset{1}{CH_2}-\!\!-COOCH_3$$

Among the various mechanisms which the IG proposes for the occurrence of m/e 143 and m/e 115, $i.$ $e.$ $(M - R - CH_2 - CH_3)^+$ where R is either a n-hexyl or a n-octyl radical, one finds exactly the same mechanism as that proposed by Djerassi. The detailed mechanism is depicted in Scheme 2. It yields m/e 143 when the 6-membered 'long-lived' transition state operates

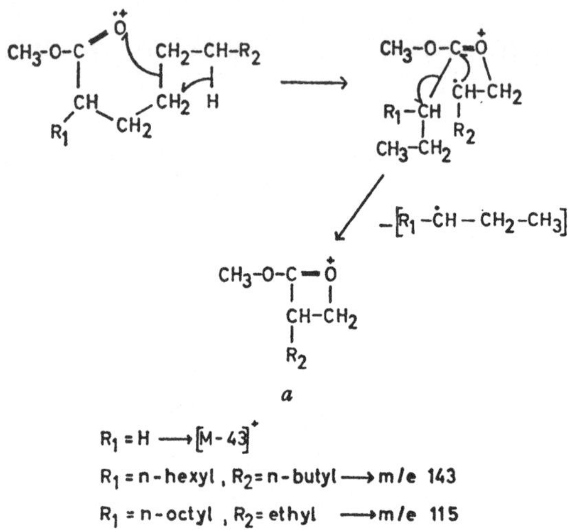

$R_1 = H \longrightarrow [M-43]^+$

$R_1 = n$-hexyl, $R_2 = n$-butyl $\longrightarrow m/e$ 143

$R_1 = n$-octyl, $R_2 = $ ethyl $\longrightarrow m/e$ 115

Scheme 2

on the n-hexyl radical connected to atom no. 9, and m/e 115 in the case where the fragmentation starts with the same cyclic transition state, but on the octyl radical attached to carbon atom no. 9. For the fragmentation leading to m/e 143, the output of the program can be summarized by the following shorthand notation:

ION 19   6LMRTR85-7   4SMRTH-5   HOM 9,10
4MRC 7,10   IMASS 143   NMASS 127   INT 30%

This sequence of operations reads as follows:

'Ionization on atom no. 19, Longlived 6-Membered Ring Transfer of a Radical with mass 85 abstracted from atom no. 7, Short-lived 4-Membered Ring Transfer of a Hydrogen atom abstracted from atom no. 5, Homolysis between atom no. 9 and atom no. 10, 4-Membered Ring Closure between atom no. 7 and atom no. 10'.

The structure of the resulting ionic species (structure a of Scheme 2) is printed in the following way:

Ionic species

O(+)—=CH2—CH—-CH2—CH2—CH2—CH3C= 1—O—CH3*

As can be seen, one of the bonds in the ring is repeated twice; in this case it is the double bond, which is followed by the symbol 1 (the first non-hydrogen atom in the output string) when it is printed for the second time.

The mechanism leading to m/e 115 is:

ION 19   6LMRTR57-14   4SMRTH-16   HOM 9,10
4MRC 16,10   IMASS 115   NMASS 155   INT 28%

In this case the transferred radical is a n-butyl radical.

C. Future Work with the Ion Generator Program

It is, of course, too early to predict all that might be accomplished with the IG program in the future. We have, however, some rather distinct preferences regarding what should be done which we would like to summarize here. First of all, the program, to be useful, must be able to work in the area of secondary ions, tertiary ions, etc., down to some lower limit on mass which would be set by the user. Hence, the IG should keep working on ions resulting from larger ions until this lower limit is reached.

Another area of interest lies in the comparison of sets of spectra of related compounds. Since the steps forming the various mechanism depend exclusively on the structures for their motivation, the types of primitive operations applied to a structure to yield mechanisms are sufficient for the purpose of comparing mechanisms from one structure to another. For example, simple β-cleavage (homolysis) followed by formation of a new bond (makebond), is the same mechanism in 5 as it is in 6.

$$H_2\overset{+}{N}-CH_2-CH_2-R \qquad\qquad R-CH_2-\overset{\overset{\bullet+}{O}}{\underset{\|}{C}}-O-R$$

$$5 \qquad\qquad\qquad\qquad 6$$

Also, the McLafferty rearrangement (the sequence of operations: short-lived 6-membered transition state with transfer of a hydrogen atom, followed by β-cleavage) is the mechanism in 7 as well as in 8, a daughter ion of 7.

$$R_1-\overset{H}{\underset{|}{CH}}-CH_2-CH_2-\overset{\overset{\bullet+}{O}}{\underset{\|}{C}}-(CH_2)_3-R_2 \qquad\qquad \overset{+}{HO}=\overset{\overset{\bullet}{CH_2}}{\underset{|}{C}}-CH_2-CH_2-\overset{H}{\underset{|}{CH}}-R_2$$

$$7 \qquad\qquad\qquad\qquad\qquad 8$$

So the need is for a program that treats the output of the IG in such a way as to identify the common mechanisms which operate in a set of compounds to yield the important ions. This could lead to the generalization of certain electron book-keeping steps as ion-formation rationalizations.

Lastly, a program could perhaps be written to find the empirical constants needed to predict mass spectra. When a mass spectrometrist fragments a structure by the method of electron book-keeping, he has some rather definite ideas about the relative intensities that some of the ions so formed should exhibit. Writing a program to do this is by no means a short-range goal, but it is an enticing accomplishment toward which to work.

IV. References

1) Hites, R. A., Biemann, K.: Anal. Chem. *40*, 1217 (1968).
2) Sweeley, C. C., Ray, B. D., Wood, W. I., Holland, J. F., Krichevsky, M. I.: Anal. Chem. *42*, 1505 (1970).
3) Smith, D. H., Olsen, R. W., Walls, F. C., Burlingame, A. L.: Anal. Chem. *43*, 1796 (1971).
4) Henneberg, D., Schomberg, G.: In: Advances in mass spectrometry (ed. A. Quayle), Vol. 5, p. 605. London: Institute of Petroleum 1971.
5) Reynolds, W. E., Bacon, V. A., Bridges, J. C., Coburn, T. C., Halpern, B., Lederberg, J., Levinthal, E. C., Steed, E., Tucker, R. B.: Anal. Chem. *42*, 1122 (1970).
6) Venkataraghavan, R.: In: Mass spectrometry. Techniques and applications (ed. G. W. A. Milne), p. 1. New York: Wiley-Interscience 1971 (and references cited therein).
7) Desiderio, D. M.: In: Mass spectrometry. Techniques and applications (ed. G. W. A. Milne), p. 11. New York: Wiley-Interscience 1971 (and references cited therein).
8) Habfast, K.: In: Advances in mass spectrometry (ed. E. Kendrick), Vol. 4, p. 3. London: The Institute of Petroleum 1968.
9) Venkataraghavan, R., Board, R. D., Klimowsky, R., Amy, J. W., McLafferty, F. W.: In: Advances in mass spectrometry, Vol. 4, p. 65. London: The Institute of Petroleum 1968.
10) Wards, S. D.: In: Mass spectrometry (ed. D. H. Williams), p. 253. London: The Chemical Society 1971.
11) Burlingame, A. L.: In: Recent developments in mass spectroscopy (ed. K. Ogata and T. Hayakawa), p. 104; Procs. Int. Conf. Kyoto, Japan, 8—12 Sept. 1969. Baltimore—London—Tokyo: Univ. Park Press 1970.
12) Holmes, W. F., Holland, W. H., Parker, J. A.: Anal. Chem. *43*, 1806 (1971).
13) Hertz, H. S., Hites, R. A., Biemann, K.: Anal. Chem. *43*, 681 (1971) (and references cited therein).
14) Erni, F., Clerc, J. T.: Helv. Chim. Acta *55*, 489 (1972) (and references cited therein).
15) Jurs, P. C., Kowalski, B. R., Isenhour, T. L.: Anal. Chem. *41*, 21 (1969).
16) Jurs, P. C., Kowalski, B. R., Isenhour, T. L., Reilley, C. N.: Anal. Chem. *41*, 690 (1969).
17) Jurs, P. C., Kowalski, B. R., Isenhour, T. L.: Anal. Chem. *41*, 695 (1969).
18) Kowalski, B. R., Jurs, P. C., Isenhour, T. L., Reilley, C. N.: Anal. Chem. *41*, 1945 (1969).

19) Jurs, P. C., Kowalski, B. R., Isenhour, T. L., Reilley, C. N.: Anal. Chem. *41*, 1949 (1969).
20) Jurs, P. C.: Anal. Chem. *42*, 1633 (1970).
21) Jurs, P. C.: Anal. Chem. *43*, 22 (1971).
22) Jurs, P. C.: Appl. Spectry. *25*, 483 (1971).
23) Jurs, P. C.: Anal. Chem. *43*, 1812 (1971).
24) Petterson, B., Ryhage, R.: Anal. Chem. *39*, 790 (1967).
25) Petterson, B., Ryhage, R.: Arkiv Kemi *26*, 293 (1967).
26) Biemann, K., McMurray, W., Fennessey, P. V.: Tetrahedron Letters *1966*, 3997.
27) Biemann, K., Cone, C., Webster, B. R., Arsenault, G. P.: J. Am. Chem. Soc. *88*, 5598 (1966).
28) Senn, M., Venkataraghavan, R., McLafferty, F. W.: J. Am. Chem. Soc. *88*, 5592 (1966).
29) Barber, M., Powers, P., Wallington, M. J., Wolstenholme, W. A.: Nature *212*, 784 (1966).
30) Biemann, K., Fennessey, P. V.: Chimia (Aarau) *21*, 266 (1967).
31) Fennessey, P. V.: Ph. D. thesis, Massachusetts Institute of Technology. Cambridge, Mass. 1968.
32) Crawford, L. R., Morrison, J. D.: Anal. Chem. *43*, 1790 (1971).
33) Crawford, L. R., Morrison, J. D.: Anal. Chem. *41*, 994 (1969).
34) Crawford, L. R., Morrison, J. D.: Anal. Chem. *40*, 1469 (1968).
35) McLafferty, Mass spectral correlations. Washington, D. C.: American Chemical Society 1963.
36) Smith, D. H.: Anal. Chem. *44*, 536 (1972).
37) Venkataraghavan, R., McLafferty, F. W., Van Lear, G. E.: Org. Mass Spectrom. *2*, 1 (1969).
38) Biemann, K., McMurray, W.: Tetrahedron Letters *1965*, 647.
39) Lederberg, J.: In: Topology of molecules, p. 37, The mathematical sciences. Cambridge, Mass.—London: MIT Press, 1969.
40) Lederberg, J., Sutherland, G. L., Buchanan, B. G., Feigenbaum, E. A., Robertson, A. V., Duffield, A. M., Djerassi, C.: J. Am. Chem. Soc. *91*, 2973 (1969).
41) Duffield, A. M., Robertson, A. V., Djerassi, C., Buchanan, B. G., Sutherland, G. L., Feigenbaum, E. A., Lederberg, J.: J. Am. Chem. Soc. *91*, 2977 (1969).
42) Schroll, G., Duffield, A. M., Djerassi, C., Buchanan, B. G., Sutherland, G. L., Feigenbaum, E. A., Lederberg, J.: J. Am. Chem. Soc. *91*, 7440 (1969).
43) Buchs, A., Duffield, A. M., Schroll, G., Djerassi, C., Delfino, A. B., Buchanan, B. G., Sutherland, G. L., Feigenbaum, E. A., Lederberg, J.: J. Am. Chem. Soc. *92*, 6831 (1970).
44) Buchs, A., Delfino, A. B., Duffield, A. M., Djerassi, C., Buchanan, B. G., Feigenbaum, Lederberg, J.: Helv. Chim. Acta *53*, 1394 (1970).
45) Buchanan, B. G., Feigenbaum, E. A., Lederberg, J.: Stanford Artificial Intelligence Project, Memo AIM-145, Report No. CS-221 (1971).
46) Delfino, A. B., Buchs, A.: Helv. Chim. Acta *55*, 2017 (1972).
47) Spiteller, G., Spiteller-Friedmann, M., Houriet, R.: Monatsh. Chem. *97*, 121 (1966).
48) Budzikiewicz, H., Djerassi, C., Williams, D. H.: In: Mass spectrometry of organic compounds, p. 179. San-Francisco: Holden-Day, Inc. 1967.
49) Budzikiewicz, H., Djerassi, C., Williams, D. H.: In: Mass spectrometry of organic compounds, p. 180. San-Francisco: Holden-Day, Inc. 1967.

Received June 26, 1972

Computer-Gas Chromatography

Dr. Friedrich Caesar

BASF, Ammoniaklaboratorium, Ludwigshafen

Contents

I. Introduction

Since the first paper in 1952[1] gas chromatography has rapidly developed and has become an important and precise tool in analytical chemistry. Early chromatograms were recorded on a strip-chart recorder and the data was reduced manually. As operational techniques were simplified, usage increased and the need for automated data reduction became apparent.

Advances in electronics and computer technology have brought about general automation of analytical apparatus. Beside this, gas chromatography has several features which favor the combination of gas chromatographs with automatic computing devices and which are responsible for the rapidly growing usage of gas chromatography-computer systems.

a) Gas chromatography is often used for routine analysis with numerous repetitions of similar samples.

b) Usually several instruments are located together in a laboratory so that several gas chromatographs can be connected to one computer.

c) Interpretation of gas chromatograms by manual techniques is very time-consuming and therefore becomes more expensive with inflation and rising wages.

d) Improvements in accuracy and precision are only possible with automatic peak-area allocation, since the manual techniques – triangulation, cut and weight, or planimetry – imply a relative standard deviation of at least 1 to 3%.

e) Gas chromatography being widely employed the development of hardware and software specially designed for use with this method has been profitable for the manufacturers, because the systems have found many customers.

The development from the first ball integrators to modern digital integrators and computer systems has been described by several authors[2,3]. The first commercial on-line gas chromatography (gc)-computer systems were combinations of several integrators with a small digital computer [4,5]. Although these systems had little trouble with data sampling and therefore worked very well, they were almost outrun by pure computer systems. The reason for this development lies mainly in the limited ability of early integrators to correctly evaluate peak areas in nonideal chromatograms. Area allocation has been improved for small peaks on the tailing of large ones by the tangent technique, and recently by a new peak-detection logic [6] which is based on the increase of area with time instead of the slope of the curve as formerly used. Baseline is not assumed to be horizontal but is computed by connecting peak beginning and end with a straight line.

Combinations of these sophisticated integrators with a computer, however, are still too expensive to compete successfully with pure computer systems. The integrating computer [7] and similar systems – *i.e.* PEP–1 [8] – may be seen as a device intermediate between a hybrid system and pure computer system, which combines the advantageous features of both.

General review articles on computer gas chromatography have been written by Perone [9,10] and Gill [11]. Most of the important papers on computer gas chromatography were presented at various symposia and were published in the December 1969, December 1971, January 1970, and January 1972 issues of J. Chromatog. Sci. and in Chromatographia *5*, 63–211 (1972). Other important lectures are contained in the Proceedings of the Seventh and Eighth International Symposia on Gas Chromatography held in Copenhagen in 1968 and in Dublin in 1970 [12].

This article will discuss several important aspects in computer gas chromatography from the viewpoint of the customer.

II. Configuration of Gas Chromatography-Computer Systems

A large number of different gas chromatography-computer systems has been described [3] for automatic processing of gas chromatographic signals which differ from each other in one fundamental feature, *i.e.* the kind of data acquisition.

A. Hybrid Systems

In an integrator-computer system or hybrid system the signal of the gc is continuously integrated by a digital integrator which transforms the original signal into peak areas and retention times. These reduced data are then used by the computer to calculate the analytical result. The accuracy and the standard deviation of the qualitative and quantitative results depend upon the features of the integrator. Modern instruments give very small standard deviations with completely separated peaks and a straight baseline; however, strongly overlapping peaks and a rising baseline as in temperature-programmed runs, are beyond their capability. Several hybrid systems have been described in the literature [4,5,13].

B. Pure Digital Computer Systems

In pure digital computer systems the voltage is sampled from the voltage/time curve of the gc at equal time intervals. After digitizing the analog voltage, the computer calculates the areas and retention times of the peaks.

With sophisticated software, overlapping peaks can be separated with high accuracy and a drifting baseline can be corrected more easily than with an integrator. The chief handicap of these systems is a great sensitivity to noise which means that peak detection tends to be unreliable. Furthermore, the standard deviation of the area calculations is generally greater than with an integrator; this is specially true with overlapping peaks. With further progress in the electronics of the gc and the development of better hardware and software, the standard deviations of area calculations should be further improved.

C. Computing Integrator

A new development [7] – the computing integrator – combines the integrating and smoothing data acquisition of the integrator with the advantages of the computer in separating the peaks and correcting the baseline.

A different division can be made between on-line and off-line systems. In off-line systems the raw data are sampled, sometimes reduced and stored on tape. In a second pass, separated in time and space from the first one, the stored data are read and digested by the computer. With an on-line system no manual handling is necessary between data acquisition and printout of the result.

D. Off-line Systems

Two types of off-line systems have been described which differ in the point at which the data flow is interrupted.

1. An integrator transforms the signals of the gas chromatograph to retention times and peak areas and punches these data on paper tape. A paper-tape reader feeds the data into the computer or into the terminal of a time-sharing computer which has the usual computing routines at its disposal. After a call to the nearest computer station, the terminal is ready to accept data and shortly after the end of the input the result is type-written as a protocol. The time-sharing service, which is widespread, particularly in the USA, offers the user all the advantages of a big computer's storage and computing capacity. The usage of a terminal requires no knowledge of computer languages. It is very cheap, since apart from rent for the terminal only the input and computing time and the storage of tables and other data must be paid for.

The role of time-sharing in gas chromatography has been recently reviewed by Gill [14]; several customers have published their experience in practice [15-17].

In general, the use of an off-line time-sharing service is limited to reduced gc data by the slow input rate and speed of transmission to the computer. For special applications it may be preferable to digitize the signal with a digital voltmeter, store the raw data on tape [18], and feed them to a computer which performs the operations specified by the software. This method may be preferable for laboratories with only a few gas chromatographs but having access to medium or large digital computer.

Desk-top computers, such as Olivetti-Underwood Programma 101 [19] or Diehl Combitron–S 10/10 [20], are also used for off-line calculations with peak areas and retention times punched on paper tape. The main problem in working with these calculators is their limited storage capacity, so that with the Olivetti 101 only a normalization procedure is possible and the data must be entered twice.

2. The time-voltage curve of a gas chromatogram can be frequency-modulated stored on tape and then, after demodulation, fed off-line at high speed into either a hybrid or pure computer system. In this pseudo on-line modification the computer works in the same way as with a rapid on-line chromatogram. Because of the limited dynamic range of the analog frequency converter, the signal must be attenuated once or several times. This system is specially suitable for testing peak allocation routines, since the stored gas chromatogram can be computed with different parameters as often as necessary. Beside the older combination of Infotronics [21], the system of L. T. T. [22] which uses an easily adaptable cassette tape recorder should be suitable for combination with a computer.

E. On-line Systems

Several types of on-line system have been described which differ from each other in size, power, structure and function.

1. Dedicated Systems

The smallest unit of an on-line system is the dedicated instrument where one computing device is combined with one instrument. For economic reasons this type is only be found as a hybrid system with desk calculators such as Diehl Combitron S [20] or Hewlett-Packard 9810A [23].

2. Dedicated Computers

Up to 64 gas chromatographs are simultaneously connected (in time-sharing) to a small or medium computer whose hardware and software is specially designed for gas chromatography. The voltage of the instruments is successively sampled by a multiplexer and stored in the core. To save storage capacity, the raw data are not stored until the end of the run but are processed continuously or in small blocks to reduced peak data such as peak areas and retention times, or peak maxima and minima and inflection points. Numerous systems of this kind are commercially available. Since the well-tested software is usually written in assembler, modifications of these programs are generally not recommended although fine tuning via adjustment of constants is frequently done. In hybrid systems in which the area determination is accomplished by the integrator the computer may be a small one; the use of desk calculators – such as Wang 700 [20] or Hewlett-Packard 9810 A [23] has also been described.

3. Mixed Computer Systems

In mixed systems several different analytical methods are connected in time-sharing with a big universal computer whose size and equipment makes it possible to program in Fortran. Using Fortran as computer language guarantees flexibility in respect of new methods and improvements. The coexistence of several different data-acquisition and computing routines in the core, however, may impede the programming of the systems and often gives rise to interference with the routines or other trouble, particularly when new or modified programs have been added. Several systems of this kind have been described [24,25,26]; they are not currently commercially available.

4. Satellite Systems

Several analytical instruments and methods are connected to a small computer – the satellite – which samples the raw data, gathers and concentrates them if necessary, and transmits them to the central computer for further processing. In this technique the total task is divided among different hardware so that each portion of the job is performed by the section best suited to do it. The satellite is completely available for data acquisition, which must be done in real time, whereas the big storage and computing capacity of the central computer enables complex calculations to be carried out.

Satellite systems have been described by Klopfenstein [27] and Günzler [21] who compared different configurations [28].

III. Hardware

The hardware of a computer system for gas chromatography consists of several units which can be discussed separately: the computer, the interface between computer and gas chromatograph, the input/output device and the gc itself.

The interface and the input device will be described in detail, as their structure determines the functioning and convenience of the system.

A. The Interface

The most important and difficult part of the system is the interface which has to fulfil four functions:

> amplify the analog signal of the gc;
> transport the signal to the computer;
> multiplex it with other instruments;
> convert it into digital form so that the digital computer can accept it.

This can be done by several methods in different sequences. A recent summary of this problem is published by Busch [29]. In hybrid systems there are virtually no difficulties with the interface. Each integrator combined with a gc produces peak areas and retention times which are transmitted and multiplexed to the computer in a digital form. Therefore, we shall discuss here only the functions of the interface in pure digital computer systems.

1. Analog-to-digital Conversion of the Signal [29,30]

The analog-to-digital conversion of the signal is generally accomplished by one of three methods:

a) voltage-to-frequency conversion,
b) the dual-slope technique, or
c) successive approximation.

a) An electronic digital integrator transforms the analog signal of the gc into a frequency and continuously accumulates the counts from the beginning to the end of a peak. In computer systems based on voltage-to-frequency conversion the counts are accumulated for a prescibed time, *i.e.* the sampling time, and then transferred to the computer [7,8b]. The advantage of the accumulation is its integrating effect which is insensitive to spikes and has good noise suppression. The frequency can be transmitted over several hundred meters without noise pick-up but conversion speed is slow for low-level input signals. Hence this method is applicable only for single-channel conversion where a converter is dedicated to each chromatograph, or with a multiplexer having few connections [31].

b) *Dual-slope technique.* This medium-speed analog-digital conversion is used for single-channel measurements as well as in multiplexer configurations up to 20 gc. Basically, this converter performs an analog integration during a fixed time by charging a capacitor. The time required to discharge the capacitor through a constant current is the digital equivalent of the input voltage amplitude [32]. It is relatively insensitive to noise because it is integrating the signal for a short time.

c) *Successive approximation.* The signal voltage is compared with the output of a digital-analog converter which is approximated to the voltage as closely as possible. This non-integrating high-speed conversion is suitable for multiplexer configurations with more than 30 gc's. Sampling rates of more than 100 kHz are possible. The sampling time of only a few microseconds increases sensitivity to noise and spikes.

2. Amplification of the Signal

a) The chief problem in the digitizing process is the need to obtain a dynamic range of 10^6 and more. The only conversion method covering this range with acceptable linearity and accuracy is voltage-to-frequency conversion [29]. For better resolution with low voltage signals, selectable gain amplifiers controlled by the computer may be used [8b,31].

b) An analog-digital converter with 20 bits would give a resolution of $1:10^6$ but it would be slow and expensive; furthermore, the converted data will be greater than one word on most computers. Depending on the system and its use, there are several configurations with ranging devices at the computer or at the gc.

Two or more amplifiers with fixed gain factors are connected in parallel to each gc. The signals of all outputs are multiplexed to the analog-digital converter and the computer selects the highest value within the dynamic range of the analog-digital converter. Several amplifiers and multiplexer connections are necessary for each gc. It is also important to take care of different drifting and varying gain ratios of the amplifiers [33]. High-voltage transmission and the use of slow amplifiers are the chief advantages of this arrangement.

c) Most multi-channel systems use an autoranging amplifier which is located between multiplexer and analog-digital converter [29]. Gain changes are controlled by a feedback gain-control loop of the amplifier or by the software of the computer. The amplifier must be fast enough to follow the cycles of the multiplexer and to avoid "talk-over" from other channels.

d) In another configuration a slower amplifier may be dedicated to each chromatograph. Though this concept is expensive for many channels, it may sometimes be preferable to transmit the signals at a high voltage level.

e) Mixed systems with one or two fixed amplifiers at the gc and an autoranging amplifier at the computer combine the advantages of high-voltage transmission with low costs [31,34,35].

f) In routine applications the attenuator at the gc may be controlled by time data stored in the computer [36] as part of a "method", or by a special module dedicated to each chromatograph [37].

3. Transmission of the Signal

The signal of the gas chromatograph may be transmitted to the computer in either digital or analog form. This depends on the location of the analog-digital converter in the line.

a) *Digital Data Transmission*

In systems with digital transmission each gc must have a device which transforms the analog signal into a digital one. The transfer rate depends on the type of analog-digital converter used. With a voltage-to-frequency (V/F) conversion a pulse rate of 1 to 10^6 pulses/sec can be transmitted by RF cables without difficulty. In V/F converters with a gated counter the data can be reduced to about 300 bits/sec and normal shielded, twisted-pair cables with integrated line-driver circuitry may be used [29]. In hybrid systems with integrators as digitizers the data rate is much reduced, since only retention times and peak areas are transmitted. In general, these data are transmitted in BCD-code sequentially via four lines or serially via $4n$ lines, where n is the number of decimal figures.

b) *Analog Data Transmission*

Noise is the greatest problem in analog data transmission. The distance between gc and computer may be several hundred meters, and magnetic and electrical fields and heavy electrical equipment switching on and off can cause trouble with low-voltage transmission. The noise picked up on the line may be stronger than the signal itself.

Several procedures are described to overcome these difficulties. The most important point is to avoid grounding loops such as arise when two instruments which are connected by a signal line are grounded with non-zero impedance and resistance. A magnetic field through the loop formed by the signal cable and the ground connections will induce a voltage occuring chiefly across the greatest resistance in this loop – the amplifier.

There are some precautions for avoiding grounding loops.

a) Computer and all gc's are connected to the same ground, usually the computer ground. This involves insulating all connections of the gc against ground, especially the mains voltage and the gas supply capillaries.

b) The ground is multiplexed with the analog line to the analog-digital converter. The shielded analog-digital converter is momentarily connected to the gc whose signal is being multiplexed. The digital output of the analog-digital converter is then transferred to the computer via a photoelectric device [32].

c) Baudisch [38] installed a separate transformer to supply power through a single line to the computer and all gas chromatographs. Likewise he provided reliable grounding with an individual ground line to each gc and the computer. He further used a double-shielded analog line.

d) Recently Schlereth and Greiner [39] described a method for transmitting low voltage over a distance up to 15 km without picking up additional noise on the line. The signal is transmitted over a standard telephone line in a current mode, compensating for additional noise by a control circuit. This device may solve many problems with gc's connected to a computer by long-distance analog lines.

In the Datacon DP 90 system [36], instead of radially connecting the gc to the computer, as is done in most systems, a data highway is used. This is a single multicore cable which connects the signal and grounding lines of each gc via a local interface to the computer. These interfaces replace the multiplexer of the radial system because they respond to computer commands to switch analog data onto the highway in sequence, as required under program control.

The output of a gas chromatograph working with a computer should have as little noise as possible. Most gc's must be cleaned up electrically, but the various authors differ in their opinion as to the work which is absolutely necessary. The best way of cleaning up is to prevent noise at its origin, since each filtering distorts the true signal. The main line has to be magnetically disconnected from the AC voltage of the FID and the output of the amplifier, which should be connected directly to the computer. The firing controls of the column oven and the injection ports in particular tend to cause spikes of high voltage and frequency. Coupling with the attenuator can produce additional noise. If there is serious disturbance of the main line, it may be necessary to install a separate transformer [35,38]. Additional hardware filtering is done after each gc. The filter, however, must have a very sharp cutoff to avoid distortion of narrow peaks and yet give a high rejection to mains and higher-frequency noise. Fozard [40] wrote that it was "painstaking work to produce a differential fourpole Butterworth filter with 3 dB down point at 5 Hz and 80 dB noise rejection at 50 Hz".

B. The Input/Output Devices

Communication between the user and the computer should be as simple and reliable as possible. Several input and output devices may be used; which of them is attached to the system is a question of philosophy and cost and depends on the configuration of the gc-computer system.

All systems have local stations next to each gc with at least two push buttons for start and stop and some lamps indicating the status of the channel.

The digital input data may be divided into three parts which are read into the computer by different pieces of equipment:

a) Information necessary to instruct the computer about the operating system and the gc software.

b) Information about the type of analysis to be performed. This is summarized in a "method", *e.g.* scanning rate, filter parameters, peak separation methods, time windows, and so on. Since these data do not change with each analysis they are generally stored in the memory.

c) Information about the sample to be analyzed. Here the data change from run to run (*e.g.* sample number and weight, number of method to be used).

a) The program and the software for the operating system are usually fed in by a card reader or a paper-tape reader in dedicated systems, where the kind and sequence of commands generally do not vary.

b) Teletypes, a paper-tape reader, or in large systems a card reader, are employed to build up a method. Small systems of limited storage capacity where methods are frequently changed are usually equipped with a tape reader. It is easy to handle but it is rather troublesome to vary a single parameter of the method. To simplify the input procedure with teletypes, it is often operated by the computer in a conversational mode with the operator. The computer asks for each parameter in the correct sequence. Illogical input data are rejected and the question is repeated. Single parameters of a stored method can be varied.

c) The input of data which are different for each run is done by means of a teletype or control panel. Small dedicated systems are usually equipped with a control panel and a paper-tape reader and larger systems with teletypes. Several larger systems, however, have control panels[26,35]. The output is written with a slow build-in printer, a teletype, or high-speed printer as used in large systems. For on-line systems it is recommended that separate teletypes be provided for input and output procedures. With a cathode viewer screen, chromatograms may be displayed and curve deconvolution methods controlled [26].

IV. Software

The software developed for gas chromatography computation consists of two almost independent parts:

A. the determination of peak areas and retention times,

B. calculations with these reduced data and the writing of a protocol.

The software for running the computer, sampling the raw data, buffering and organizing them in the core or on the disc will not be discussed here.

In hybrid systems the most critical part of the software is replaced by the integrator which produces peak areas and retention times. Repeatability and accuracy of the results are given by the properties of the integrator, which are usually well defined by the manufacturers.

A. The Determination of Peak Areas and Retention Times

Several operations are necessary for calculating retention times and peak areas from the time-voltage curve of the gc:

1. Sampling the raw data,
2. Smoothing the raw data,
3. Peak detection and integration,
4. Determination of the baseline,
5. Separation of overlapped peaks.

For some elements of the software it is essential for the choice of method to know whether the raw data have to be computed in real time to the sampling, or whether they can be stored until the run is complete.

1. Sampling the Raw Data

Sampling has to be frequent enough to measure the area of the peak as accurately as possible. A greater sampling rate than necessary does not improve reproducibility but impedes detection of the beginning and end of the peak and merely overloads the computer.

Theoretically, 13 points per peak [41] are required to evaluate the area within 0.1% error by integration according to Simpson, assuming resolved gaussian peak shape. This is true also for the retention time as interpolated by a quadratic equation. Baumann [42] et al. reported that only 7 to 8 scans per peak are necessary for 99.99% area recovery of a gaussian peak by numerical integration using the trapezoidal rule. They neglected the problem of peak detection and noise. Since there is always noise and often asymmetric and overlapping peaks, 20 to 25 points across a peak should be considered a realistic figure [40,43]. P. Sutre and J. P. Malengé [44] investigated the relative error of an isolated gaussian peak as a function of the signal/noise ratio and the number of points per peak. They found that 100 points/peak are necessary to reach a relative error of 10^{-3} with a signal/noise ratio of twenty. These values correspond to the results of Chesler and Cram [45] who found that with a signal/noise ratio of 1 000 – as produced by a 10-bit analog-digital converter with maximal signal – up to 100 points/peak are necessary for 99.978% recovery of a gaussian peak. For peaks with strong tailing, the area recovery is only 99.8% since the tailing end disappears in the noise. The standard deviations for the above conditions are 0.02 and 0.1% respectively.

The reliability of peak detection based on the first and second derivatives depends on the slope of the curve, *i.e.* on the number of points/peak. In isothermal chromatograms the broadening of the peak with retention time must be taken into consideration by means of hardware or software manipulations. With sample-point reduction during the run, broad peaks with a small slope will be transferred to narrow peaks with more distinct first and second derivatives. This may be achieved by varying the sampling rate in steps controlled by retention time stored in the computer or by condensing several samples into one medium point.

The sampling rate for each channel may be synchronized either with a quartz crystal or to the mains power frequency [47,48]. A quartz crystal guarantees that the samples will have exactly the same time difference so that peak integration is independent of mains frequency variations. Small differences between the sampling rate and power frequency, however, produce a low-frequency oscillation superimposed on the gc signal with an amplitude equal to the amount of mains frequency noise, and this cannot be eliminated by either hardware or software filters. When the multiplexer rate is an integral multiple of the power frequency and the sampling is synchronized with the main frequency, the sampling occurs at the same point on the power-frequency waveform. This not only avoids the low-frequency oscillation but also eliminates most of the mains noise. Variations in mains frequency, however, then have a great influence on the accuracy of area assignment.

2. Digital Smoothing the Raw Data

Although hardware filters eliminate much of the noise, especially of high-frequency noise, software smoothing of the samples is necessary to guarantee exact detection of the beginning and end of the peak. Integration of the peak area, however, may also be done with unsmoothed data since integrating itself has a smoothing effect [31].

Several procedures have been published and may be used alone or in combination with each other.

a) *Bunching or Bracketing*

The bunching routine takes several successive readings from the analog-digital converter and combines them into one output value. This procedure – also called group averaging – performs two functions: it provides smoothing and also enhances the sensitivity of the peak detection logic. It can only be used if more points than necessary have been sampled across the peak.

b) *Exponential Filtering*

Exponential filtering, a fast procedure, is equivalent to RC-type analog filtering. It must be used with great care, because it involves a lag in the curve which can cause errors in the case of overlapping peaks. The general filtering formula is [44]

$$y_{t \text{ filt.}} = \frac{y_{t \text{ unfilt.}}}{2^k} + \left(1 - \frac{1}{2^k}\right) y_{t-1 \text{ filt.}}$$

c) *Moving Average*

This procedure calculates the average of several input readings before and after the point to be smoothed. After adding the next input reading and dropping the last one, a new average is calculated. The values are weighed equally in a linear moving average or may be supplied with different positive or negative factors. The general formula is

$$V_{t \text{ smoothed}} = \sum_{i=0}^{i=i} \frac{K_i(V_{t-i} + V_{t+i})}{2 \sum\limits_{i=0}^{i=i} K_i}$$

d) *Least-Square Curve Fitting*

This procedure, which has been fully described by Savitzky and Golay [50], fits a quadratic or cubic curve by the least-square criterion to a definite number of input points using the convolution-constant method. The center point is then replaced with the corresponding point of the curve. After moving one data point forward the process is repeated. Though very effective, this smoothing is tedious even with a high-speed computer. As shown by Savitzky and Golay [46,50] the procedure is exactly equivalent to a moving average with different coefficients for each term. The value of these factors depends on the degree of the convolution curve and the number of points involved in the calculation. This is only valid with equidistantly spaced points; if this is not effected by the sampling rate, the data have to be interpolated to equal time intervals [51]. The first and second derivatives of the peaks are also determined by a moving-average procedure using unprocessed or smoothed data [49]. The coefficients K_i again depend on the degree of the polynomial and the number of points used [50,52].

The more smoothing is done, the more distorted a peak becomes, but with careful application these routines make it possible to smooth noisy data considerably. Data to which too rigorous smoothing criteria have been applied can be extremely reproducible but quite meaningless [40]. The influence of smoothing parameters on the raw data with exponential filtering [44], linear moving average [31] and least-square fitting [40,42,31] has been investigated by several authors.

In gas chromatography-computer systems exponential filtering [44,53] or a least-square fit [54,55] may be used alone, or bracketing and a linear moving average can be combined with a least-square curve fit [7,32,55]. A method of exponential filtering combined with a nine-point least-square fit [26] has also been published.

3. Peak Detection and Integration

The detection of characteristic peak data, such as beginning, maximum and end of a peak, the valley between two peaks, inflection points and tangent points on shoulders, is based on a combination of the first and second derivatives, or is done with the first derivative alone. In the PEP-1 system peak-detection logic is based on the increase in the integral with time [8c]. A simple peak-detection logic using only the absolute signal has been derived by Fozard [56]. Peak detection by both the second and the first derivatives has advantages as well as handicaps. Peak detection by the second derivative is most sensitive to noise but independent of baseline drift. The first derivative, though less disturbed by noise, cannot determine the peak data necessary for area calculations of shoulders and for nonlinear separation of strongly overlapping peaks.

Since noise cannot be wholly avoided, two types of parameters are important in detection logic:

a) deadbands or thresholds, *i.e.* the interval at around zero where the derivatives y' and y'' can vary without it being considered as a peak (noise on the baseline), and

b) number of confirmations: y' and y'' must have a certain number of successive values outside the deadband for a peak to be considered to have begun or ended.

Integration of a peak is simply performed by summing the height of each sample between peak beginning and end, multiplied by the time interval between successive samples, or by considering a peak as a group of trapezoids. Also integration according to Simpson is used where three successive points are connected by a parabola. This may be done with either smoothed or unprocessed data since integration itself is a smoothing procedure. As a peak has already begun when the detection logic becomes positive, integration of the area generally starts from several samples before the point at which the detection was confirmed [7,8b,42]. The areas of overlapping peaks can be obtained by integration of the analytical function whose parameters were derived by a peak-deconvolution program [43,52]. The total area of the overlapping peak group obtained by numerical integration has to be correlated with the sum of the integrated derived functions [51,57].

4. Determination of the Baseline

Baseline determination is one of the most difficult and important tasks in peak-area allocation, since the location of the baseline greatly influences the value of the areas. Except for some simple algorithms, there are few details in the literature about baseline construction.

A straight line may be plotted between the start of the first and the end of the last peak. If intermediate points are located below this line the program takes this into consideration by plotting a broken line passing through these points [26]. In most software packages the segments of the curve confirmed by the peak detection logic to be the baseline are connected with straight lines, with due consideration of the points of the curve below this line. In order to improve baseline calculations in difficult temperature-programmed chromatograms, a valley between fused peaks can be selected by time to be a forced baseline point [8c]. Baseline segments may also be connected with a polynomial of second or third order. Although more accurate than a straight line, this is most sensitive to noise. Impressive success in baseline construction has been demonstrated by Schomburg [49].

As early as 1967 Simon [59)] published a baseline computation on the basis of a series of extremely strong moving averages. This method, which seemed very promising, does not appear to have been used, perhaps because all raw data of the entire run are necessary for the procedure.

5. Separation of Overlapping Peaks

A still unresolved problem in gc is the allocation of the areas of overlapping peaks with the same precision as areas of isolated peaks. Since the relative standard deviation of gc is in practice often better than 1% and optimally better than 0.1%, it is desirable to use peak-area determinations having smaller systematic errors and deviations than 0.1% for overlapping peaks, too. It is still a challenge to mathematicians to seek general algorithms which satisfy the above conditions.

The principal difficulty in calculating overlapping peak areas is that the time-voltage or time-current curve produced by the detector cannot be described by a simple mathematic model and − worse − that the curve is not reproducible from run to run. It varies with the amount of the substance [60,61)], the skill of the technician in injecting, the temperature and absorption/desorption phenomena in the column (tailing). The peak shape is also affected by the kind and amount of the neighboring peaks, especially in the overlapping region, since during the whole separation process in the column the various compounds influence each other in the gas phase as well as in the stationary phase. Without proof, it is always taken for granted that the detector indicates a mixture of two or three compounds eluting together out of the column as the sum of the single compounds.

Hence it is not surprising that existing peak-resolving programs are only able to approximate the calculated areas to the true values. Two kinds of peak-separation methods are used:

a) linear methods with good repeatability but often large systematic errors and

b) nonlinear methods with usually poor standard deviation but higher accuracy.

a) *Linear Peak-Separation Methods*

Linear peak-separation methods are the basis of most separation programs in gc-computer software. The computation costs little computer time so that they can be used in real time with data sampling.

The simplest method is to drop a perpendicular from the valley between two peaks on to the baseline. Several authors have investigated the systematic errors of this method with two overlapping gaussian peaks as a function of the resolution factor and their height relative to each other [62-66]. The errors may be compensated for by correction factors. In practice, however, the application of these corrections based on the gaussian peak shape is of limited value [65] because almost all peaks show tailing and this becomes more significant with increasing peak size. Only if a small peak precedes a larger one can correction factors improve the result [65].

Fused peaks of widely different amplitude may be separated by a tangent from the valley between the peaks to the end of the smaller one. This method, called skimming, is useful with small peaks on the tailing of large ones or on a solvent peak but very misleading with fused gaussian peaks [63]. Skimming can be defined for shoulders in the same way as for peaks, replacing the valley by the inflection point.

Skimming and dropping a perpendicular form the basis of all commercial gc-computer programs; the decision as to which method should be employed is based on the relative height of the peaks or on a time reference defined in the method.

This combination is best suited for small computer systems with real-time calculations, since only a few characteristic points must be stored for the peak separation. These points are defined by the first and second derivatives (maxima, minima and point of inflection) or can be easily determined during real-time processing (tangent point or slice point).

Discontinuity in changing from perpendicular to skimming is avoided by the "democratic distribution" mentioned by Günzler [21]. A tangent is drawn from the valley between two peaks to both edges. The area formed with the baseline is distributed to the peaks over the tangents in proportion to their areas. This method is appropriate for small peaks on the tailing of large ones but gives rise to large systematic errors with overlapping gaussian peaks. Unlike skimming, it cannot be used in the real-time mode since, to draw the tangent backwards, the raw data of the leading edge are required at the time when the valley between the two peaks is determined.

Some other possible methods of linear peak separation have been published, but they are not used in computer gas chromatography.

A triangular model, described by Westerberg, is not only less accurate but also less reproducible than dropping a perpendicular [66]. Several rules outlined by Kaiser and Klier [67] for drawing a separating line between unresolved peaks give large errors with realistic peaks [65].

The true height of one peak in a group of two or three overlapping peaks is obtained by subtracting the contributions of the other two [68]. A general feature of all linear separation methods is their sensitivity to the degree of overlap and height ratio so that correction factors which may be experimentally determined for compensating the systematic errors should be a function of height ratio and resolution.

157

b) *Nonlinear Separation Methods*

In contrast to the calculations described above, nonlinear separation methods try to allocate the area of overlapping peaks by taking into account the true peak shape as far as possible.

Overlapping peaks can be quantitatively analyzed by constructing mathematically a synthetic chromatogram which is fitted to the experimental one by iterative, nonlinear regression analysis. This method, called curve fitting, requires two assumptions.

First, it is necessary to know the general shape of the peaks to be fitted even though the shape is not relevant in obtaining the analytical result. Second, one must have an initial estimate of the peak parameter to be determined. In the case of badly overlapping peaks, these parameters cannot be taken automatically from the experimental chromatogram. With poor initial estimates, however, the analysis may not converge or may result in a convergence on to incorrect values.

Several authors have described their experience in resolving synthetically overlapping and real gas chromatographic peaks by the curve-fitting procedure.

In the simplest case a gaussian peak shape is assumed with three parameters (H, t_0 and w (height and time of the maximum and width of the peak)). The general formula is

$$y(t) = H \cdot e^{-\left(\frac{t-t_0}{w}\right)^2}$$

The asymmetry of real peaks may be accounted for by the two different widths for the leading and the trailing halves of the peak [43,63] which implies the calculation of four parameters per peak. Gladney [51], Roberts [58], and Littlewood et al. [69] used a combination of gaussian distribution and exponential lag to take into account the tailing of real peaks. Littlewood pointed out that from theoretical considerations it is reasonable to expect that experimental gc peaks might approximate very closely to the exponential gaussian convolute shape. This combination enables area analysis of overlapping peaks to be made more accurately than when pure gaussian peaks are fitted. For simplification, he reduced the number of parameters by assuming that within a peak group the time constant of the exponential delay is constant and the peak width is proportional to the retention time.

When no exact analytical forms of the peaks are known, an experimental peak shape can be used for curve fitting. The only condition is that the equation of the peak may be written [66,70]

$$y(t) = H \cdot f\{(t - t_R), w\}$$

When the peak shape f is obtained in a separate experiment in which a single peak is digitized, the parameters (H, t and w) of the overlapping peaks are found by fitting them to the experimental curve.

The existing technique of deconvolution usually breaks down when the overlap is so strong that the composite appears as a single peak, *i.e.* below the detectibility limit. This problem can be dealt with by employing statistical-moment [71-73] or slope analysis, as described by Grushka *et al.* In the former, variations in the peak shape, as obtained from the skew and excess, are used as indicators in the recognition between a true single peak and a composite made up of two peaks in the form of a single band. In the latter, deviations in the behavior of the second derivative indicate the existence of double peaks. These analyses are applied to pure gaussian and exponentially modified gaussian peaks.

So far now these methods have not been used in practical computer gas chromatography, although the statistical moments can be readily calculated from the points of inflection [74].

The main features of curve fitting are:

1) For regression analysis the curves to be analysed must be overdetermined. Usually three or four points per parameter to be determined are required, *i.e.* nine to twelve points for a gaussian peak. On the other hand, software smoothing need not be done since curve fitting is itself a smoothing procedure.

2) Since all raw data of a peak group have to be stored until the computation begins, curve fitting cannot be done in the real-time mode. These methods require a complex program which is usually run off-line or as background work.

With these handicaps in mind, several manufacturers of gc-computer systems have developed nonlinear separation methods designed to be used in the time domain. These methods are based on the fact that a gaussian peak is defined by three or four points which assume a symmetrical or asymmetrical shape.

With given shape factors for the leading and trailing halves of the peak and ignoring the displacement in time between the maxima of the true peaks and the corresponding maxima in the fused waveform, Hancock and co-workers calculated the true heights A_i of the resolved peaks in real time [63]. The peak area is then given by

$$F_i = \tfrac{1}{2} A_i (C_1 + C_i) \sqrt{\pi}$$

The only data taken from the chromatogram are time and heights of the peak maxima of the overlapped curve.

Roberts [75] described a "Short-Cut Fused-Peak Resolution Method" for gaussian peaks which requires two points per peak and an initial estimate for the height and the width which can be taken from the experimental curve. Assuming the time of the maxima to be fixed and known as observed in the trace, the heights and widths of the resolved peaks are fitted to the sampled points by several iterations.

Time and height of the minima and maxima of the experimental curve are used by Metzger [57] to allocate areas of overlapping peaks in real time. The height and the two shape factors are obtained in a first approximation and can be improved by repeating the calculation and considering the time difference between the maxima of the true peaks and in the unresolved curve. The only additional information the operator has to supply is the ratio of the shape factors for the leading and trailing halves of the bi-gaussian peaks. The peak areas obtained by numerical integration and dropping a perpendicular are then corrected only in the overlapping regions with the areas calculated from the derived gaussian function. This procedure is much more accurate than integration of the complete peak areas via the gaussian function. It guarantees that the sum of the resolved peak areas is correct. In an IBM program [76] special points of the chromatogram are approximated to an analytical or experimental peak form. The points known from the first and second derivatives are the maxima, minima, inflections and slice points of the envelope. The peak forms are defined by a set of 13 points which can be fed from punched cards or computed by a shape-fitting program during a standardization run.

6. Testing the Precision and Accuracy of the Computer System

The most important feature of a gc-computer system is the precision and accuracy of the analyses obtainable with it. Whereas formerly errors were caused by manual area determination, the gas chromatograph itself is now the limiting factor. The quality of the area allocation and the time measurement of the computer system are masked by the variations from run to run of the chromatograph itself, although the precision of gas chromatographic analysis with computer data handling can be better than 0.02% rel. with large resolved peaks [86,87]. Baumann [77] was the first to test the repeatability of the computer system independently of the repeatability of the chromatograph by simultanously connecting two channels of the computer system to one gas chromatograph. Fozard [40] modified this experiment by feeding three channels with a high DC voltage while five other channels were progressively attenuated. With this method the linearity of the system and the reproducibility between the channels can be proved and "talk-over" between the channels be detected. Another possibility is to feed a chromatogram, frequency modulated and stored on tape, into the computer several times.

The accuracy of peak deconvolution programs can be tested with synthetically overlapping peaks [43,58]. Another procedure is useful for checking deconvolution of experimental peaks: a gas chromatograph is supplied with two gas sample loops of different volume connected sequentially in the carrier gas line. Sampling with both loops at different time intervals enables each combination of overlapping identical peaks to be produced. Landowne [78] and Hancock [63] moved the peaks into each other by raising the temperature of the column oven. In the two last experiments the variations of the gas chromatography have not been eliminated, but the errors of the peak deconvolution methods are still larger than those of the gas chromatography.

If all digitized samples of a chromatogram can be stored [25,26], it is possible to examine the influence of smoothing and peak separation parameters on the calculated peak areas. This is important for optimizing the values of these parameters.

B. Calculations with Peak Areas and Retention Times

1. Qualitative Identifications

For quantitative interpretations of the chromatogram the detected compounds must be identified, and response factors have to be applied to transform peak areas into weights. Automatic identification of peaks may be accomplished by numbering the peaks, on the basis of absolute retention time bands or by retention times relative to an internal time standard. Since retention times relative to one standard are not constant enough for reliable identification [77,79], especially in long chromatograms, multistandard systems are used which correct the deviations in time from one standard to the next [32,53]. The best realization of the multistandard principle is the identification with Kováts Retention Indices applicable in isothermal [80] and temperature programmed runs [81,82]. Any compound with a known retention index may be used as a standard in isothermal [25b] as well as in temperature-programmed chromatograms [81]. It may be claimed that automatic identification in chromatograms with many compounds is only possible with Kováts indices or related procedures.

The software of many gc-computer systems includes the calculation of Kováts indices; a program in Fortran has been published by Castello et al.[83]. Schomburg [25b,84] has described the advantages of automatic index calculations; with effective thin-coated capillaries standard deviations of 0.05 units are attainable. Kováts indices are also suited for identification in low-temperature-programmed gas chromatography with packed columns [85], where standard deviations with nonpolar compounds over a short time are better than 0.3 units [81].

2. Area Calculations

Programs for calculations with area values are trivial for computer software. The extent of their availability, which is essential for the convenience and utility of the system, depends on the configuration of the computer system and on the capacity of the core and external stores.

The following subprograms are part of most software packages:

a) 100% area standardization,
b) (100−x)% area standardization,
c) analysis with internal standard,
d) analysis with external standard;

all methods may be used with or without response factors.

Other subprograms have been designed according to the wishes of the user:

e) grouping of several peaks to one area value,
f) simulated distillation [88],
g) data collating [88]. In this program the results of several runs on different columns are digested and combined into one report. Summary reports of several runs may be obtained giving only information of special interest. The results of each run are stored in the computer [89] or on a magnetic-tape cassette recorder interfaced to a dedicated gc data system [90].

In many medium and large computer systems gas chromatographic data may be stored for a certain time after the run has been completed and the result printed. This allows the user to modify the computing process without repeating the gas chromatographic run. These stored data may be:

1) digitized raw sample points,
2) characteristic peak data, such as beginning, maximum and end of a peak, valleys, points of inflection, and slice points,
3) peak areas and retention times.

Whereas the first type is important for fine-tuning the processing of the raw data, the second and third types are useful for correcting peak identification, improving the printout of the protocol, and applying other subprograms to modify the original information. Schomburg [25] and Guichard [26] have described in detail the possibilities and convenience of recalculating a chromatogram on the basis of raw data or of a "basic report".

V. Computer Control of the Gas Chromatograph

A computer directly connected with a gas chromatograph is able to accept and digest signals from the instrument, to send instructions to the instrument and to control their execution. These commands may be based on a time scale stored in the computer or given in response to an incoming signal. Welland and Muir [91] have discussed the principles of computer-controlled operation of a gas chromatograph.

As a first stage of control, the setting of instrument parameters at certain preset points of time is possible with many gc-computer systems [36,49,76,91]. These parameters may be: point of injection (when an autosampler is used), end of run, attenuator switching, backflush, start of temperature program, column oven open and closed, column switching, and so on. In hybrid systems the settings of the integrator may also be changed automatically [20].

The second stage of control is to check the execution of the commands transmitted to the instrument. This has been used in a "watchdog" capacity by Lyons [92] for supervision of a preparative gas chromatograph during overnight operation.

The third stage of control is to measure the result of instrument settings, *e.g.* gas flow and temperature. The instrument can then be held after new settings have been made until gas flow and temperature are stable [91,93]. It may also be used to check the baseline before sampling [92].

The most advanced stage is closed-loop control, where the output of the instrument determines new parameter settings. Thurman, Mueller and Burke [94] described an automatic experiment for calculating the HEPT as a function of temperature and gas flow. The new settings of gas flow and temperature are computed from the last HEPT calculations, then sent to the instrument and their implementation checked. As soon as gas flow and temperature are stable enough, the new run begins with automatic injection. Temperature and gas flow are held stable by the instrument's hardware, not by the computer. Only the very exact temperature and gas flow readings are included in the HEPT calculation.

For automatic routine analysis in a manufacturing plant, closed-loop control operations may be favorable. A gas chromatograph with autosampler could repeat runs of a single sample until the mean of the runs is reliable enough for a decision. The computer would compare several runs and then decide whether the process needed to be modified or whether the next sample could be analyzed.

Recently Gill [2] stated that very little work has been published in this field but that this is considered to be a very important area in future development.

VI. Economics

In considering the economics of a gc-computer system, the cost of purchase or rent must be set against the savings and advantages expected from using the computer. While it is relatively easy to calculate the sum spent or saved by the computer, it is very difficult in many areas to estimate the advantages of the system and to express these benefits in money terms.

These advantages are a) faster analysis time, b) increased accuracy and smaller standard deviation in quantitative analysis, c) precise determination of retention data with Kováts indices in qualitative analysis, d) increased utility of instruments, e) improvement and application of new techniques which cannot be used without a computer, and f) more information overall. Each of these improvements is a step forward in the progress of analytical science and technology, some of which, however, will earn their keep only in the future.

How much does an analytical result with a small standard deviation contribute toward optimizing a process or selling a product? A process can only be optimized if the standard deviation of the analytical result is smaller than the variations produced in the process when one factor of the reaction conditions is varied.

If a product is to be sold with a guaranteed purity of 80%, the content must be at least 82.5% pure, measured with five determinations at an absolute standard deviation of 2% [95]. With a standard deviation of 0.5%, a purity of only 80.6% is necessary. With a production of 1000 tons per year at a price of $.25/kg, the value of the more exact analysis will be $4750 per year.

In routine analysis, there is no doubt that a computer system pays for itself in time saved by technicians who would otherwise perform the analyses by manual techniques. This is especially true of smaller systems and time-sharing services. Jackson has calculated the economics of gc automation via an integrator/time-sharing combination [16]. Schroeder and Walter [95] described a saving of more than one dollar per sample and operating costs of less than $.80 per sample. Other authors [15,17] reported the cost of computation per analysis to be about $1.50 for 200 analyses per month. Lyons [97] compared the technician time required for various manual operations with the time necessary for working with integrators, and off-line and on-line computer systems. The treatment of an analysis with 10 peaks and 14 min elution time rquires 50 min with triangulation and 10 min with an on-line computer. He estimated pay-back time for an on-line computer working with twenty gas chromatographs to be two years; this can be compared with a life expectancy of at least five years. With a large mixed computer system and 6,000 samples per month, Price [98] reported that he saved nine men minus the personel necessary to maintain the system in operation. Schomburg [25a,49] calculated a total cost of $1.75 per chromatogram with the mixed computer system in Mühlheim, assuming that the computer handled 15,000 samples per year. This calculation allows for the fact that only 5% of computer system time is devoted to chromatographic work. The cost per chromatogram with a commercial dedicated system amounts – on the basis of 15,000 analyses/year – to $ 1.73. From this comparison Schomburg [49] concludes that preference for one or the other system cannot be justified on the basis of cost per chromatogram alone.

VII. Trends in Future

Some future trends in instrumentation and the economics of computer-gas chromatography seem fairly clear [3]. New electronic techniques will bring down the cost of digital computers and make them faster, more efficient, and more reliable. On the other hand, wages will continue to rise so that the use of computers in gas chromatography will become more and more attractive. This is especially true of small systems. Already manufacturers can replace specialized parts of the software with read-only memories (ROM), a new hardware device. This hardware memory can be read on a nanosecond time scale, but it cannot be altered by the user. On the other hand, the cost of producing software will rise with wages. Therefore the expense of generating and running big systems with a large amount of versatile software will decrease only slowly. The general trend is thus toward more hardware in smaller systems. In future routine applications each chromatograph may be supplied with a hardware device which will process the signal in the same way as nowadays a small dedicated system or an integrating computer.

Acknowledgements. I wish to thank the Badische Anilin- & Soda-Fabrik AG for permission to publish the paper.

Literature

[1] James, A. T., Martin, A. J. P.: Biochem. J. *50*, 679 (1952).
[2] Gill, J. M.: J. Chromatog. Sci. *10*, 1 (1972).
[3] Anderson, R. E.: Chromatog. *5*, 105 (1972).
[4] Cotton, J. M.: In: Progress in Anal. Chem., Vol. 4. New York: Plenum Press 1970.
[5] Peterson, G. V., Zerenner, E., Eccles, M., Kapuskar, W.: In: Anal. Advan. Hewlett-Packard Autumn 1970.
[6] SIP-1, a paper of Perkin-Elmer *9*, 1972.
[7] Hettinger, J. D., Hubbard, J. R., Gill, J. M., Miller, L. A.: J. Chromatog. Sci. *9*, 710 (1971).
[8] a) Paul, G. T., Sloughter, W. J.
   b) Gill, H. A., Lee, R. E., Condon, R. D.
   c) Noonan, D. J., Condon, R. D.
   d) March, E. W., Pieper, E. C.
   All presented at the Pittsburgh Conference on Anal. Chem. and Appl. Spectr., Cleveland, Ohio 1971.
[9] Perone, S. P.: J. Chromatog. Sci. *7*, 714 (1969).
[10] — Anal. Chem. *43*, 1288 (1971).
[11] Gill, J. M.: J. Chromatog. Sci. *7*, 731 (1969).
[12] Gas Chromatog. 1968 and Gas Chromatog. 1970, Proceed. of the Seventh and Eighth

F. Caesar

Sympos. on Gas Chromatog., published by The Institute of Petroleum, London: 61 New Cavendish Street.

[13] Malan, E., Brink, B.: Chromatog. *5*, 182 (1972).

[14] Gill, J. M., Henselman, J.: Chromatog. *5*, 108 (1972).

[15] Tuinstra, L. G. M. Th., de Graaff, J. B. H. D.: Chromatog. *4*, 468 (1971).

[16] Jackson, H. W.: J. Chromatog. Sci. *9*, 706 (1971).

[17a] Tochner, M., Magnuson, J. A., Sonderman, L. Z.: J. Chromatog. Sci. *7*, 740 (1969).

[17b] Rathgeb, K.: Chromatog. *4*, 270 (1971).

[18] Hegedus, L. L., Petersen, E. E.: J. Chromatog. Sci. *9*, 551 (1971).

[19] Walker, B. L.: Anal. Biochem. *37*, 44 (1970).

[20] Bischet, G., Knechtel, G.: Chromatog. *5*, 166 (1972).

[21] Günzler, H.: Chem.-Ing.-Tech. *42*, 877 (1970).

[22] Lignes Télégraphiques et Téléphoniques, Paris: Nouveau système d'interface entre chromatog. et calculateurs (1972).

[23] Hewlett-Packard, information sheet on "a versatile, on-line gc data processing system", *6*, 1972.

[24a] Ziegler, E., Henneberg, D., Schomburg, G.: Anal. Chem. *42*, (9), 51 A (1970).

[24b] — — — Angew. Chem. *84*, 371 (1972); Angew. Chem. Internat. Edit. *11*, 348 (1972).

[25a] Schomburg, G., Weeke, F., Weimann, B., Ziegler, E.: J. Chromatog. Sci. *9*, 735 (1971).

[25b] — — — — Angew. Chem. *84*, 390 (1972); Angew. Chem. Internat. Edit. *11*, 366 (1972).

[26] Guichard, N., Sicard, G.: Chromatog. *5*, 83 (1972).

[27] Klopfenstein, C. E.: J. Chromatog. Sci. *10*, 22 (1972).

[28] Günzler, H.: Z. Anal. Chem. *256*, 14 (1971).

[29] Busch, U.: Chromatog. *5*, 63 (1972).

[30] Gäumann, T.: Chimia *25*, 144 (1971).

[31] Charrier, G., Dupuis, M. C., Merlivat, J. C., Pons, J., Sigelle, R.: Chromatog. *5*, 119 (1972).

[32] Varian Aerograph: Chromatog. Dat. Syst. Oct. 12, 1969.

[33] Siemens AG: Programmsystem PRAG 1972.

[34] Metzger, H. D., Radszuweit, K. G.: Chromatog. *5*, 186 (1972).

[35] Papendick, H. D.: Erdöl u. Kohle *24*, 637 (1971).

[36] D. P. 90 Datacon: Philips Electronic Industrie GmbH, 1970.

[37] Deans, D. R.: J. Chromatog. Sci. *9*, 729 (1971).

[38] Baudisch, J.: Chromatog. *5*, 79 (1972).

[39] Schlereth, G. A., Greiner, K.: Chromatog. *5*, 70 (1972).

[40] Fozard, A., Franses, J. J., Wyatt, A.: Chromatog. *5*, 377 (1972).

[41] Kishimoto K., Musha, S.: J. Chromatog. Sci. *9*, 608 (1971).

[42] Baumann, F., Herlicska, E., Brown, A. C., Blesch, J.: J. Chromatog. Sci. *7*, 680 (1969).

[43] Hock, F.: Chromatog. *2*, 334 (1969).

[44] Sutre, P., Malengé, J. P.: Chromatog. *5*, 141 (1972).

[45] Chesler, S. N., Cram, S. P.: Anal. Chem. *43*, 1922 (1971).

[46] Steinier, J., Termonia, Y., Deltour, J.: Anal. Chem. *44*, 1906 (1972).

[47] Infotronics: Model CRS—1000.

[48] Honeywell: GC—16, Gas Chromatog. Package.

[49] Schomburg, G., Ziegler, E.: Chromatog. *5*, 96 (1972).

[50] Savitzky A., Golay, M. J. E.: Anal. Chem. *36*, 1627 (1964).

[51] Gladney, H. M., Dowden, B. F., Swalen, J. D.: Anal. Chem. *41*, 883 (1969).

[52] Hancock, H. A., Lichtenstein, I.: J. Chromatog. Sci. *7*, 290 (1969).

53) Wehling, I.: Chromatog. *5*, 197 (1972).
54) Baan, A.: Graduation Report, Eindhoven University of Technology.
55) Datachrom One, System description from Instem Ltd., Stone.
56) Fozard, A., Franses, J. J., Wyatt, A. J.: Chromatog. *5*, 130 (1972).
57) Metzger, H. D.: Chromatog. *3*, 64 (1970).
58) Roberts, S. M., Wilkinson, D. H., Walker, L. R.: Anal. Chem. *42*, 886 (1970).
59) Simon, W., Castelli, W. P., Rutstein, D. D.: J. Gas Chromatog. *5*, 578 (1967).
60) Baudisch, J., Papendick, H. P., Schlöder, V.: Chromatog. *3*, 469 (1970).
61) Mc Nair, H. M., Cooke, W. M.: J. Chromatog. Sci. *10*, 27 (1972).
62) Kishimoto, K., Migauchi, H., Musha, S.: J. Chromatog. Sci. *10*, 220 (1972).
63) Hancock jr., H. A., Dahm, L. A., Muldoon, J. F.: J. Chromatog. Sci. *8*, 57 (1970).
64) Proksch, E., Bruneder, H., Granzner, V.: J. Chromatog. Sci. *7*, 473 (1969).
65) Novak, J., Petrovic, K., Wičar, S.: J. Chromatog. *55*, 221 (1971).
66) Westerburg, A. W.: Anal. Chem. *41*, 1770 (1969).
67) Kaiser, R., Klier, M.: Chromatog. *2*, 559 (1969).
68) Mori, Y.: J. Chromatog. *66*, 9 (1972); *70*, 31 (1972).
69) Anderson, A. H., Gibb, T. C., Littlewood, A. B.: J. Chromatog. Sci. *8*, 640 (1970).
70) — — — Anal. Chem. *42*, 434 (1970).
71) Grushka, E., Meyers, M. N., Giddings, J. C.: Anal. Chem. *42*, 21 (1970).
72) — Monacelli, G. C.: Anal. Chem. *44*, 484 (1972).
73) — Anal. Chem. *44*, 1733 (1972).
74) Grubner, O.: Anal. Chem. *43*, 1934 (1971).
75) Roberts, S. M.: Anal. Chem. *44*, 502 (1972).
76) 1800 Chromatog. Monitoring Program (5718—XX1), IBM, 1970.
77) Baumann, F., Brown, A. C., Mitchell, M. B.: J. Chromatog. Sci. *8*, 8 (1970).
78) Landowne, R. A., Morosani, R. W., Herrmann, R. A., King jr., R. M., Schnus, H. G.: Anal. Chem. *44*, 1961 (1972).
79) Raymond, A. J., Lawrey, D. M. G., Mayer, T. J.: J. Chromatog. Sci. *8*, 1 (1970).
80) Caesar, F.: Chromatog. *5*, 173 (1972).
81) — unpublished results.
82) Jaeschke, A., Rohrschneider, L.: Chromatog. *5*, 333 (1972).
83) Castello, G., Parodi, P.: Chromatog. *4*, 147 (1971).
84) Schomburg, G.: Chromatog. *4*, 286 (1971).
85) Kaiser, R.: Chromatog. *5*, 117 (1972).
86) Obtained with a F 30—PEP-1 system with autosampler by Perkin–Elmer. Private Communication from R. Kaiser, Bad Dürkheim.
87) Varian Aerograph, information sheet "The Complete Chromatography Data System".
88) Hendrickson, L.: In: Research Notes, Varian, Sept. 1972.
89) Cracen, D. A., Everett, E. S., Rubel, M.: J. Chromatog. Sci. *9*, 541 (1971).
90) Frazer, J. T., Guran, B. T.: J. Chromatog. Sci. *9*, 718 (1971).
91) Welland, J. M., Muir, A. R.: Chromatog. *5*, 136 (1972).
92) Lyons, J. G.: Chromatog. *5*, 156 (1972).
93) Swingle, R. S., Rogers, L. B.: Anal. Chem. *43*, 810 (1971).
94) Thurman, R. G., Mueller, K. A., Burke, M. F.: J. Chromatog. Sci. *9*, 77 (1971).
95) Calculated after Kaiser, R., Gottschalk, G.: Elementare Tests zur Beurteilung von Meßdaten, Bibliographisches Institut, Mannheim/Wien/Zürich 1972.
96) Schroeder, D. L., Walther, H. W.: J. Chromatog. Sci. *10*, 14 (1972).
97) Lyons, J. G.: Gas Chromatog. 1970, Proceedings of the Eighth Symposium on Gas Chromatog., London: published by The Institute of Petroleum, 61 New Cavendish Street, page 302.
98) Price, J. G. W., Scott, J. C., Wheeler, L. O.: J. Chromatog. Sci. *9*, 722 (1971).

Received January 8, 1973

Computer-Assisted Instruction in Chemistry

Wolfgang Geist and Dipl.-Ing. Peter Ripota

Projekt CUU (Computer-unterstützter Unterricht), Universität Freiburg

Contents

I. Introduction

In general, computer programs are devised to solve complex numerical problems, to process large quantities of data in standard ways, or to control instruments. But computers can also be employed as teaching aids. There are two basic approaches:

1. programs can be written to simulate processes
2. programs can be applied to the mathematical treatment of experimental data.

An instructor can be highly successful in teaching a small number of students. But today the number of students is increasing and no great efforts are being made to increase teaching facilities and qualified teachers. The subject matter is becoming more and more complex and the instructor is faced with additional responsibilities, so that the student often plays a passive role without active participation in the instructional process.

One expedient would be to use less experienced or temporary tutors to provide more individual attention in courses of instruction. However, this creates financial problems; moreover, there is a lack of qualified instructors.

The learning situation would be ideal if each student had a tutor of his own. This tutor should be able to introduce new concepts, to provide help, hints, periodic reviews, and encouragement, to determine when the student is in trouble, to make sure that he understands, and to tell him what is wrong with his answers.

The desired individual attention can be provided by employing *computers* as teaching devices. Such a system can release an instructor from many functions which do not need his active participation. There is a body of information to be learned by the student which requires very little or no teaching by an instructor. The computer is a patient tutor to guide the student through the material to be learned. The ability of the computer to interpret the student's answers, and to make decisions about a student on the basis of data distinguishes it from other well-known instructional media (*e.g.* scrambled books).

The computer can perform the following tasks:

1. Provide the student with study guidance on a personal basis.
2. Set exercises simulating laboratory work. Thus the student learns to collect, manipulate, and interpret data.
3. Offer a chemical course which is to some extent individually tailored.

The computer is thus an active learning aid. The student's own decisions and actions determine his progress. He is engaged in decision-making processes and is allowed to proceed at his own pace.

Educational research has shown that there is very little transfer of what is learned, from one subject to another unless a specific effort is made to teach this. An instructor has little time to engage in such efforts. A computer has as much time as is needed. This is why people invented Computer-Assisted Instruction (CAI).

II. Computer-Assisted Instruction

Computer-Assisted Instruction (CAI) was initially conceived as a device for transferring programmed instruction from a book to the computer. In this kind of instruction, a simulated dialog takes place between the student sitting at a terminal (a typewriter or graphic display) and the computer, or more correctly the program: the computer asks questions and the student tries fo find the correct answers. If he succeeds, he gets a reward in the form of some encouraging comment. A number of teaching strategies have evolved from work on programmed instruction via the computer, among them tutorial, drill-and-practice, problem-solving, etc. Languages were developed especially for CAI, e.g. Coursewriter [22]. These provided certain useful editing features for interactively deleting, inserting and otherwise changing instructional material; they contained procedures and features for analyzing answers (these never attained the level of grammatical parsing but were limited to checking for keywords, which in most cases proved sufficient.) They were, however, unable to calculate the result of 3 divided by 2.

After some years of practice, an exciting discovery was made: computers are able to compute. New teaching strategies and new programming languages were developed for CAI. The strategies included simulation, gaming, data banks. The languages — e.g. Tutor [2] or Planit [5,20] — used real numbers and offered all the facilities of a higher programming language as regards numerical analysis while retaining the features of string processing and editing. However, some kind of dialog, deemed necessary by the authors, remained even though it was not really necessary. Few programs made use of powerful interactive systems like Basic or APL for presenting and solving numerical problems.

This is the current situation. To take a glimpse into the future, we think that both students and teachers will become more computer-conscious. There should be a greater awareness of the facilities offered by the computer for providing data (both numerical and non-numerical), checking calculations, simulating natural phenomena, and running laboratory exercises in conjunction with conventional instructional methods (among which we count CAI in its older form). There are new programming languages designed for interactive use of the central processor, which are both simple in nomenclature and powerful in execution. Communication with the computer will take place in more 'natural' surroundings — natural for the computer, if not for the student. After all, for a human-like dialog the best partner is another human being and not some contrivance of transistors, however large their number.

However, there is no doubt, that CAI will always be *"supplemental* to rather than a *supplantation* of the human teacher" (Lagowski [24]).

The literature on CAI is scattered over a number of technical reports, papers presented at conferences, a very few good articles, and even fewer books. For a list of readings, see Refs. 1, 13, 14, 18, 19, 21, 43.

CAI, reasonably applied, has certain advantages over conventional instruction. These advantages are neatly summarized by Lower [27]:

1. The student receives the full attention of the instructor (the computer) during the entire instructional period.
2. The student has control of the selection of course material.
3. He has control of his rate of progress during the program.
4. The student is no longer forced to attend at an arbitrarily scheduled period once or twice a week (CAI terminals are usually available all day, even at weekends).
5. The student is an *active* participant in the instructional session; this ensures a degree of personal involvement that is not always present during lecture and class sessions.

A. Tutorial Programs

Basically, there are three teaching strategies for instructional programs:

> Tutorial programs
> Drill-and-practice programs
> Problem-solving programs

In a *tutorial*, the dialog usually consists of questions asked by the program, helpful hints for wrong (but intelligent) answers, remedial sequences when the student's knowledge (or lack of it) fails to conform to the author's ideas, etc. Control is firmly in the grip of the program, and the student's choice consists of supplying the answer, asking for further information, reviewing completed sections, or skipping others if his past performance warrants it. Supplementary equipment includes audio-visual material: tapes, slides, microfiches, etc.

A *drill-and-practice* program concentrates on presenting many simple problems to encourage learning by heart of instructional material. Although there is no sharp line of division between this strategy and *problem-solving*, there is a shift of emphasis in instructional objectives: Drill-and-practice strives for learning by heart, whereas problem-solving emphasizes transfer functions, *i. e.* the student is expected to learn how to apply certain principles to the solving of specific problems, and to be able to transfer problem-solving techniques from one group of problems to another.

We now present some examples.

Many students have great difficulty in solving mathematical problems, and in some chemical courses mathematics is very important. Courses covering stoichiometry, gas laws, solutions, etc. demand mathematical skills. Therefore many instructional courses contain sections which provide students with drill-and-practice exercises in mathematics.

At Simon Fraser University in Canada [37] the Mathop program has been developed to teach elementary mathematical problems.

Here is a sample dialog:

It is possible to have negative numbers as exponents. A number with a negative exponent such as A**(—B) is simply (1/A)**B. Now what is 4**(—2) expressed as a simple fraction?

———→ 1/4

You have most likely made a sign error. Try again!

———→ —1/4

4**(—2) = 1/4**2   What is the number?

———→ 16

You have forgotten to invert the answer. Try again!

———→ 1/16

Good!

(Student input is marked by an arrow and boldface characters.)

J. D. Dye and V. A. Nicely [11] of Michigan State University (USA) have developed programs providing statistical analysis of experimental data. The programs can be used by students in a analytical-physical laboratory course. These programs fit an arbitrary function of not more than 20 parameters and four variables to a data set containing not more than 99 points. Furthermore, they can solve arbitrary algebraic equations having not more than 20 unknowns, or they can minimize (or maximize) an arbitrary functional having not more than 20 parameters. These programs run off-line only.

Courses in elementary chemistry are offered at nearly all CAI centers where chemistry is taught. Most of the programs contain all three basic instructional strategies.

At the CAI center of the University of Freiburg (W. Germany) Baumgartner, Geist, Müller and Schmidt developed a program in elementary chemistry called "CUS-Chemie", which tests the chemical knowledge of newentrants to the university [4]. If the student's knowledge is found to be deficient, he is branched into a remedial part of the program or referred to a book which will supply the missing knowledge. The program covers the following aspects:

Periodic system of elements
Structure of molecules (orbital theory)
Redox equations
Chemical bonds
Chemical kinetics
Acids and bases

Here is an example of a dialog between the program and a student:

Wodurch läßt sich prinzipiell schweflige Säure H_2SO_3 in Dithionsäure $H_2S_2O_6$ überführen?

——→ durch Reduktion

Falls Sie geraten haben, haben Sie falsch geraten. Pech!
Falls Sie aber durch Überlegung zum falschen Ergebnis kamen, hier eine Zwischenfrage:

Um die obige Frage beantworten zu können, muß Ihnen der Begriff Oxydationsstufe geläufig sein.

Welche Oxydationsstufe hat der Schwefel in H_2SO_3 und in $H_2S_2O_6$

——→ 3 bei H_2SO_3 und 4 bei $H_2S_2O_6$

Leider falsch!

Betrachten Sie die Oxydationsstufenbestimmung des Schwefels in der Schwefelsäure H_2SO_4!

$H2 = 2\,H = 2*(+1) = +2$
$O4 = 4\,O = 4*(-2) = -8$

Die Differenz zwischen der Oxydationsstufe des Sauerstoffs und des Wasserstoffs ist (-6).

Da die Summe aller Oxydationszahlen in einer Verbindung Null sein muß, hat Schwefel in H_2SO_4 die Oxydationsstufe $(+6)$.

Welche Oxydationsstufe hat demnach Schwefel in H_2SO_3 und in $H_2S_2O_6$?

——→ 4 und 5

Danke, es stimmt! . . . .

One objective of this program is to avoid difficulties of comprehension in the course "chemistry for beginners". The program was written in Coursewriter III and has been translated into Planit.

At Simon Fraser University (Canada) S. K. Lower [37] has written a program called Elementary Chemistry Problems. It has a problem-tutorial concept and the software used is Coursewriter III. Here is an example of a student-computer dialog:

> That's the TOTAL (positive) oxidation number of ALL the Fe.
>
> Now what is the oxidation number of EACH Fe atom in Fe_2O_3?
>
> ──→ 3
>
> OOPS! You forgot the SIGN! Please try again! PLEASE REPEAT!
>
> ──→ +3
>
> OK, very good — now how about FeO_4? (Notice that this is a negative ion)
>
> ──→ +3
>
> NO — you are just guessing! Try again!
>
> ──→ −8
>
> Try again, Genius! This would give you an ion with an overal charge of −16.
>
> ──→ +8
>
> NO — you are not taking the overall negative charge of the ion into account. Please try again or type   )?)
>
> ──→ ?
>
> What is the total oxidation number of oxygen in FeO_4-?
>
> ──→ −8
>
> OK, now what number must the total POSITIVE oxidation numbers in the molecule add up to?
>
> ──→ +8
>
> No, don't forget that this particular molecule is an ion — it carries two negative charges. These charges result from an imbalance of the total positive and negative oxidation numbers — the negative oxidation number is in excess.
> Please try again!
>
> ──→ +6
>
> OK, so what will be the oxidation number of Fe in FeO_4-?
> (Notice that this is a negative ion)
>
> ──→ +6
>
> OK.

Lower has also written a course called Balancing Equations which covers the techniques of balancing chemical equations, both ordinary and redox types.

S. Castleberry and J. J. Lagowski [8] of the University of Texas at Austin (USA) have devoted a great deal of effort to testing computer teaching techniques. They developed programs in remedial, drill, examination, tutorial, and simulated-experiment format. They found that there is a body of information in general chemistry which requires student practice.

Typical subjects are:

> Stoichiometry
> Concentration calculations
> Colligative properties
> Aspects of atomic structure and equilibrium
> Chemistry preskills (metric unit conversion and energy unit conversion)
> Heat, work, and energy
> Kinetic theory
> Thermochemical problems
> Relationship between formulas, mass, moles and gas volumes
> Balanced equations
> Atomic structure.

Experimental groups of students following the practice courses scored significantly higher than a control group on those portions of examinations related to the programmed material. Moreover, the experimental group was highly motivated.

At the same CAI center L. B. Rodewald and G. H. Culp and J. J. Lagowski [33] tried to use the computer to teach organic chemistry. They compiled a course using tutorial and drill-and-practice strategies. The course consists of nine instructional modules:

> Structure and geometry of alkanes
> Skeletal isomerism of alkanes and stereochemistry of cycloalkanes
> Nomenclature of alkanes and cycloalkanes
> Preparation of alkanes
> The mechanism of the chlorination of ethane
> The halogenation of alkanes
> Fundamentals of stereochemistry
> General synthesis:
> 1. Alkene-related synthesis
> 2. Electrophilic aromatic substitution-related synthesis

Another CAI center, at the University of Illinois, Urbana, is also busily engaged in developing instructional courses in elementary chemistry. There are also some programs on more difficult subjects. The software used is the Plato III system where each student has a television screen and a keyset. The Plato keyset has a number of characters specially for chemists: arrows to right ($\rightarrow$), arrows to left ($\leftarrow$) and for equilibrium ($\rightleftharpoons$). R. C. Grandey [15,16] has written a course in elementary chemistry. There are computer-aided lessons in which a student determines a chemical formula from composition by weight. He has to find out quantities from chemical equations and to balance equations for oxidation-reduction reactions. The chemical equations displayed on the screen are similar to those the student would find in a textbook.

Each lesson has three parts: introductory section; practice problems; diagnostic quizzes.

The student can work through all sections at his own speed. He can request additional information in form of help, hints, further practice, and reviews. The sequences determining decisions are made on the basis of the results of one or a series of questions.

In the practice problem section the student is given data for a specific problem for which he has to find the solution. He is allowed to ask for help or to use the computer as a calculator (most of the CAI systems have calculator capabilities, accessible at any time).

The student is able to supply his own problems in the lessons on the determination of chemical formulas and calculation of quantities from chemical equations.

Here is an example of a dialogue between the student and the computer:

Type the balanced equation for the chemical reaction:

$\longrightarrow$ **Ba(OH)$_2$ + H$_2$SO$_4$ $\rightleftharpoons$ BaSO$_4$ + 2 H$_2$O**   OK

Type the formula for the chemical species for which you are given complete quantitative data!

$\longrightarrow$ **Ba(OH)$_2$**   OK

Type the formula of the species for which you want to calculate the quantity!

$\longrightarrow$ **BaSO$_4$**   OK

PRESS NEXT

Ba(OH)$_2$ + H$_2$SO$_4$ $\rightleftharpoons$ BaSO$_4$ + 2 H$_2$O

What is the known quantity of Ba(OH)$_2$? (include units)

$\longrightarrow$ **450 g**   OK

In what units do you want to calculate the quantity of BaSO$_4$?

$\longrightarrow$ **kilograms**   OK

The computer displays the data and requests the student's answer:

Ba(OH)$_2$ + H$_2$SO$_4$ $\rightleftharpoons$ BaSO$_4$ + 2 H$_2$O

450 grams Ba(OH)$_2$

   x kilograms BaSO$_4$

How many kilograms of BaSO$_4$ are involved in the reaction with 450 grams Ba(OH)$_2$?

$\longrightarrow$ **0.613 kilograms BaSO$_4$**   OK

You can use the computer as a calculator by pressing TERM and typing calc.

177

As can be seen, the program has to match numerical answers. Here is an example for such a matching of numerical answers.

The correct answer is: 625 ± 2 grams

Student's answer	Computer's response
625	Please label your answer!
625 g	OK
624 g	OK
$6.24 * 10^2$ g	OK
628 g	Your answer is too high
621 grams	Your answer is too low
$6.24 * 10^{23}$ g	Your answer is too high

Furthermore, the program has to interpret chemical formulas. For example if the student types "NH_3OH" for the formula of ammonium hydroxide he should be told that the hydrogen is incorrect and not that he has misspelled a word. The routine EQTJUDG was written to interpret any chemical equation. The routine controls the balancing of mass and charge in chemical equations and checks other rules for writing chemical equations, *e.g.*

Chemical symbols must be valid

All parentheses must be followed by a subscript

All subscripts must be integers

All superscripts must include a $+$ or $-$ sign

All superscripts must be integers

Another routine, Search, was developed to determine the presence or absence of the formula of a particular chemical species in a chemical equation (*e.g.* ions in oxidation-reduction equations)

The equation

$$3\, H^+ + NO_3^- + 3\, e^- \rightleftharpoons NO + H_2O$$

was interpreted as incorrect. The error message was:

The following are not balanced:

H   O   charge

The program consists of the following lessons:

A. Math-Skill

B. Chemical formulas

C. Calculation of solutions and quantities from chemical equations

D. Balancing oxidation-reduction-equations

The lessons fulfill the following objectives:

In lesson B the student has to recognize the information contained in a chemical formula, calculate the formula weight from the chemical formula, and determine the molecular formula and molecular weight.

In lesson C the student must know the definition of molarity and moles, the quantitative relationship of a chemical equation to determine quantities of reactants and products for reactions. Lesson D has the following objectives: Assignment of oxidation numbers to elements according to a set of rules; balancing oxidation-reduction equations by the half-reaction method; and identification of oxidizing and reducing agents.

The student's record lists all activities and data concerning him: name, time elapsed since he started the lesson, any requests for special features or help, and the student's responses.

Typical students' comments are: "I am able to take the time I need to absorb and comprehend the material without inconveniencing another person." "One knows immediately if the answer is correct or incorrect."

S. Smith [40] working at the same CAI center devolved a course in organic chemistry.

Here is a sample dialog:

Indicate the reagents required for each of the following conversions

The student answers:

The computer indicates "OK" if the student gives the answer "Br_2". The program accepts 18 different answers in this case.

B. Problem-Solving

At many CAI-centers specific problem-solving programs have been developed for elementary chemistry or more difficult problems. Thus, R. C. Grandey [16] has developed a program in elementary chemistry which contains problem-solving sections; S. Smith has put some problem-solving sections in his organic chemistry program: the student has to interpret nmr spectra and solve simple problems in chemical synthesis and chemical analysis.

L. J. Soltzberg [42] of Simmons College, Boston, Massachusetts, has realized that there are often mathematical barriers which block any qualitative feeling for chemical processes. The student is often incapable of solving differential equations which are very important in studying kinetic systems. Wave-mechanical analysis of chemical systems is an important development which confers predictive power. The computer can give the student a qualitative feeling for the microscopic systems described by wave mechanics. The computer program evaluates the radial part of any hydrogen-like wave-function. The student has only to specify values for the nuclear charge Z, the principal quantum number n, and the azimutal quantum number l. The program computes and plots on the teletype the radial part of the wave function and the radial electron-density distribution. All programs run on a Basic software system.

The systems described so far are all on-line systems. But there is also the possibility of using off-line systems for problem-solving procedures. C. Wilkins and C. E. Klopfenstein [46] wrote programs which calculate first-order rate constants and activation parameters by the least-squares method. This is particularly useful in demonstrating how sensitive activation parameters are to small errors in experimental rate constants. The student learns that it is necessary to be precise in recording data from his own experiments.

Craig, Sherertz, Carlton, and Ackermann [10] have written some off-line problem-solving programs in Fortran IV G which extend the student's experience beyond what is feasible or even possible at any given level in the laboratory. The computer makes it possible to provide the student with wider experience in data interpretation and hypothesis-making. In quantum mechanics the computer can provide rapid and direct access to the solutions of quantum-mechanical models in the form of electron-density distributions for atomic or molecular orbitals. So the student acquires a feeling for the behavior of such systems. The Orbital program calculates electron-density-probability maps for atomic orbitals. Molorbit calculates molecular orbitals. Both are based on the LCAO approximation. Further areas to benefit from the use of such problem-solving programs are chemical kinetics and nmr spectroscopy.

The Enzyme program calculates enzyme kinetics on the basis of Michaelis-Menton mechanics. The Titrate program has been developed for use in experiments with polyprotic acids. Spin 3 is a nuclear-magnetic-resonance program which permits the student to explore the relationship between the appearance of nmr-spectra and some critical parameters such as chemical shift, coupling constants and the radio-frequency for a three-spin system. As an output the student gets typical nmr spectra.

Computers can also be used for marking individual assignments. The students can verify the results on the computer before entering them on a mark-sense card for grading. Such a problem-solving application of the computer has been developed by N. D. Yaney of the Purdue University at Hammond, Indiana (USA) [47].

Similar programs for problem-solving in undergraduate chemistry have been written by M. Bader of Moravian College, Bethlehem, Pennsylvania [3]. These programs are able to calculate nmr spectra, Hückel molecular orbitals, hydrogen-atom electron-density contour diagrams, and molecular-orbital contour diagrams.

Numerous other CAI programs in chemistry have been written at a variety of institutions in Europe and America. For a complete survey, see Lekan [26].

C. Data Banks

Data banks are information retrieval systems, providing the user on request with information on certain subjects, or with problems of a certain kind. Thus, the programs of the previous chapter may be categorized as data banks. Requests may be in natural language ("Why did Nero play the violin?") or in some kind of code ("C22—!—A1 & D23"), they may involve 'aspects' (*e.g.* in a computerized library, see "The Techniques Used to Retrieve Information and Data from the Literature" in this volume); they can be used in a batch mode or interactively, alone or imbedded in some regular course. Only the latter aspect is of interest in this article.

A good example of a data bank integrated into a regular CAI course is *ornom* (organic nomenclature), developed by Brian Funt at Simon Fraser University. A tutorial program introduces the student to the concepts and logic of organic nomenclature. Anywhere during the program he may ask the computer to 'draw' a formula for the name he (the student) supplies. If the name is composed of compatible parts, a structural formula is drawn by the program, after which the student is transferred back to the main sequence. Thus newly acquired knowledge may be checked immediately, thereby giving instant feedback and enhancing the learning process.

Here is a sample dialog from *ornom*:

What would you call   CH3–CH2–CH=CH2

———→ **draw**

Type the name of the structure you would like drawn.

———→ **2-ethyl-3-pentene**

The name you have written, 2-ethyl-3-pentene, is incorrect.
The trouble is with the multiple bond position number. It is too large a number. You may type the correct name of a structure and I will try again to draw it or you may type 'return' to get back the main program.

———→ **2-ethyl-2-pentene**

Because of the length and position of the substituents on the main chain the longest chain will be more than 5 carbons long. You may try the structure you wish drawn again, or, to get back to the main program, type ›return›

———→ **2-methyl, 2-pentene**

```
        CH3
       /
CH3–C=CH–CH2–CH3
```

OK that is it.

For another structure, type ›draw›; otherwise, answer the last question you were asked.

Example is taken from [38].)

D. Simulations and Games

As mentioned above, the general trend in CAI is away from the simple tutorial where the computer chats on happily while its partner almost never has any real freedom of word or choice, and towards a more learner-controlled environment, as realized in *simulations* and *games*. The Structuring of the material (done by the author in tutorials) is left to the student (which presupposes the student's ability to do so).

If we try a definition or *simulation*, we have to stress two aspects or equal importance [32]:

1. It is a *model* of a real system or process (*e.g.* a system of differential equations describing a chemical reaction);
2. the model is *realized* on another, usually artificial, material system, or on a computer.

The realization of the model is a process whose temporal unfolding is related to that of the original system, *i.e.* there is some kind of slow-motion or fast-motion transformation. According to this definition, the programs described in chap. B. are not simulations; they are effective algorithms for generating data of significance to chemical phenomena.

In *games*, an element of competition is added. Competition may take place among the human participants (as with political games), or it may be a confrontation of the student's abilities with the author's wits as implemented in the program.

From the point of view of teaching strategies, simulations are carefree explorations of dynamic models whose only purpose is that of playfully getting acquainted with the simulated system. Games on the other hand, put the student in a stress situation where, relying on limited resources of information and time, he has to achieve a predefined goal, and is assumed to acquire factual and even operational knowledge in the process.

Pure simulations are usually no substitute for learning scientific facts; without a hypothesis about the principles underlying the simulated process no real understanding is possible. However, when used as a supplement to a regular course (lectures, tutorial CAI programs, books etc.), both simulations and games (the latter usually simulations of laboratory exercises) may be of considerable benefit.

Now let us enumerate some situations where simulations and games can be of value[35,39,45].

1. An experiment might consume a disproportionate amount of time if performed in its entirety, or, on the other hand, a chemical process might proceed too fast to be of any observational value (*e.g.* in kinetics). (Argument of *time*)

2. An experiment might be conceptually simple but might involve apparatus too complex for beginning students to manipulate (*e.g.* atomic spectroscopy, X-ray diffraction, etc.). (Argument of *complexity*)

3. An experiment might depend on skill in manual techniques unconnected with the logic of the whole exercise, thus making it impossible for the student to concentrate on the essentials of a method, technique or process. (Argument of *logic*)

4. An experiment might involve expensive equipment which cannot be supplied for a large number of students on an individual basis. (Argument of *money*)

5. Use of physical facilities might be severely restricted because of too many students, crowded schedules, or other administrative problems. (Argument of *availability*)

6. An experiment might involve hazardous substances or equipment, rendering it unfit for inexperienced students, although it might be of chemical or educational importance. (Argument of *danger*)

And here are some of the advantages of simulations and games:

Students can become familiar with experiments they would normally be unable to perform for one of the above-mentioned reasons.

A considerable amount of space, time, and money can be saved by running simulated experiments on the computer.

Time: Lagowski [24] reports a reduction of learning time from 12h of laboratory time to 4.5 h of computer time to complete a certain qualitative analysis (60% reduction), and, in another instance, a reduction from 131.8 min to 66.7 min (50%).

Money: According to Lambe [25], the ratio between the cost of laboratory experiments and computer simulations is approximately 2:4.5, but this is a general statement about readily available and easily accessible laboratory techniques. When complex apparatus or expensive substances (*e.g.* radioactive elements) are involved, this ratio will be much more in favor of simulations.

The 'cookbook effect' is abolished. Students are forced to think for themselves in a problem-solving, decision-making learning environment.

The student is freed from the tedious task of manipulating test tubes and other uninspiring equipment, which enables him to concentrate on the essentials and the logic of a chemical process or experimental action.

Students can work with more samples than would be possible if they worked only in a chemistry laboratory.

Here are some suitable experiments for simulation:

Qualitative and quantitative analysis
(gravimetric, titrametric and instrumental techniques)
Emission spectroscopy
X-Ray diffraction
The Mullikan oil-drop experiment
The Rutherford scattering experiment
J. J. Thompson's discharge experiments
Kinetic experiments
(simulation of chemical processes:
Brownian motion and other diffusion phenomena, chemical reactions, steady-state kinetics, etc).

And here are some examples.

1. Chemical Analysis

The aims of this kind of simulation are: to provide the student with a large number of chemical substances from which he may choose as many as he likes (and his time allows); to familiarize him with the process of decision-making in chemical experiments without having him bother too much about the consequences of his actions; and to apply theoretical concepts of chemistry in a situation as similar to the real laboratory situation as possible.

A typical program contains a number of laboratory techniques and a set of compounds unknown to the student. The student chooses one technique, and is immediately shown the results of his investigation; he may then choose another technique, ask for information (tables of data etc.) or try a guess as to the nature of the unknown substance. Remedial sequences ("You should try first to...") may be interwoven for inexperienced students.

But there is one significant difference from the real world. When performing an experiment at the terminal, the student has to tell the computer what to do, whereas in the real world he does these things himself. We may think of two extremes of this kind of communication. The first, in natural language, somehow resembles a dialog between a master and his servant; the second, a highly formalized kind of communication, uses a code similar to a programming language, which makes interaction concise but also more difficult to comprehend without practice.

A good example of the first kind of program is provided by Lagowski [24] and by Smith [40], and of the second kind by Burkhardt [7]

A typical dialog of Smith's program (written in Tutor) looks like this:

   ⟶ **What colour is it?**

The solid is white and the liquid colorless.

   ⟶ **What is its melting point?**

The melting point is 19—21°.

   ⟶ **What does the nmr look like?**

This is the 60 MHz proton nmr spectrum.
For a table of chemical shifts, press DATA.
         (The spectrum is shown on the screen.)

   ⟶ **What is the unknown soluble in?**

Please provide more information.

   ⟶ **Is it soluble in H_2SO_4?**

The unknown is soluble in cold concentrated H_2SO_4.

   ⟶ **Elemental analysis.**

We can give you results from a sodium fusion. To try it, press LAB.
      (etc.)

The student may ask questions in any order and in any way he wishes; either complete sentences or simple key words may be used. If there is a word 'X' the program doesn't understand, a message is printed:

Please forgive me, the following word is not in my vocabulary: 'X'.

According to the authors, experience in rapid identification of unknown organic compounds is provided by this lesson.

On the other hand, Burkhardt's program (written in Lidia [36]) is devoid of the human touch so dear to American instructors. Here, a typical dialog looks like this:

⟶ **19**	Student chooses substance
Colorless liquid	Short description
⟶ **+**	Student asks for table of
(Display of available methods)	available methods
⟶ **9**	Student chooses one method
(Result of investigation)	
⟶ **?**	Student asks for more information
(More information)	
⟶ **Sulfuric acid**	Guess
Right. New substance?	Guess was correct.

At the end of each analysis, a list of applied techniques is printed, together with an indication as to their usefulness (*e.g.*, 'color of litmus' is meaningless with solids or gases; this is indicated by a '!' following the method).

To add an element of competition — remember that we are dealing with games — each method may be given a price corresponding to expenditure in the real world. 'Winner' of the game is the one arriving at the correct solution with the least simulated money spent (Bitzer [6]).

2. Chemical Synthesis

An interesting example of a program simulating a laboratory course in chemical synthesis is given by Smith [6] and [41]. First, the student chooses a compound, say a benzene derivative, he wants to synthetize. This may be done in the following way:

Use the NEXT key to move the small arrow to the appropriate position on the ring. Type the desired group and press NEXT. You may have one or two substituents.

When the structure is complete, press LAB.

(Student's input is printed in boldface.) As can be seen, the student choose $C_6H_4BrNO_2$ as the target product. To determine the structure of the product, the computer first classifies the starting material by type, location and number of functional groups. Now the student is presented with the problem of converting benzene to the choosen compound. He does this by indicating the reagents required in each reaction to achieve — step by step — the indicated conversion. The first step may look like this:

.. and the next (and last) step would be:

Whereupon the student gets the message:

The synthesis is completed.

After a reagent is 'added' (at the terminal) to benzene or its derivative, the program checks to determine if a reaction is possible, then assigns the new group to the most reactive site on the ring. If further substitution is not impossible, or incompatible with already attached groups, the student is told this and asked for a new input. Hints are presented upon request ("HELP"). As can be seen, this program combines aspects of simulation, gaming, tutorials and data banks in a most effective way.

III. The Computer in Laboratory

Obviously digital computers have become an indispensable part of chemical experimentation in industry as well as in the university. Therefore it is necessary for the chemist to understand the proper use of digital computers in the laboratory. There should be curricula providing background knowledge on computerized chemical instrumentation. An adequate appreciation of the new technology cannot be acquired through reading alone. There must be a training program which supplies experience with digital computers in a chemical laboratory.

S. P. Perone and J. F. Eagleston [29] of Purdue University, Lafayette, Indiana, have taken the first steps in developing a program involving the introduction of on-line computer application in the undergraduate chemistry laboratory. There is an introductory course in analytical chemistry which deals with quantitative chemical measurements. They eliminated none of the chemical and instrumental principles in the course. The objective was to incorporate the on-line computer into laboratory experiments. The student has to understand and carry out experiments as usual — but in addition he has to design appropriate programming for optimal interaction between the computer, the student, and the experiment. The student has to learn the proper way to use a computer to solve mensurasion problems in chemistry. There are three aspects:

1. The computer as a control element in experimentation.
2. Real-time interaction with experiments.
3. The computer as an integral part of chemical instrumentation.

The success of this program depends on the availability of two items:

1. A high-level programming language allowing conversational mode sequences;
2. a special hardware system.

At Purdue University Basic, a high-level programming language, was chosen because:

1. Basic is easy to learn.
2. Basic is an algebraically-oriented conversational language.
3. Basic is interactive.
4. Basic is a universally accepted language.
5. Basic is available at commercial time-sharing terminals.

So at Purdue University a software system has been developed called "Purdue Real-Time Basic (PRTB)".

The advantages of the Purdue curriculum are:

1. The student is exposed to modern computer technology in laboratory experiments.
2. The student is encouraged to use more rigorous data-processing approaches.
3. The student becomes keenly aware of the factors which limit and define experimental accuracy and precision.
4. The student's interest in quantitative chemical experimentation is stimulated.

Here are some processes simulated by the on-line computer at Purdue University:

> Kinetic methods of analysis
> Amperometric titration
> Coulometric titration
> Potentiometric titration
> Spectrometric analysis
> Fast-sweep polarographic analysis
> Thermal analysis
> Synthesis experiments
> Chromatographic experiments.

Perone and Eagleston have described an on-line computer program simulating the process of gas chromatography. This process involves a wide variety of data-processing problems. Data-processing problems in this program are, for example:

> Peak integration
> Precise peak location and identification
> Non-uniform detector responses
> Resolution of overlapping peaks
> Non-zero baselines
> Accurate area determinations
> Noisy data
> Qualitative identification
> etc.

Another application of computers in the laboratory is as *plotters*. To facilitate the construction and manipulation of a three-dimensional molecular model, Portugal and Minicozzi [31] have designed and written algorithms which transform topological and geometrical information into dynamic molecular displays. Written in Fortran IV, the program permits:

> 1. Scaling
> 2. Horizontal and vertical translation
> 3. Rotation around an axis
> 4. Display of the molecule with or without its hydrogen atoms
> etc.

The picture is displayed in perspective with atom centers and bonds represented as points and lines.

The same goal, visualizing mathematical concepts, is pursued by Peterson and Fuller [30] who use digital computers as a means for calculating electron-density probabilities and supplying a graphic output. Similar programs were devised by Bader [3] and Ewig, Geig and Harris [12]. The latter group utilize an interactive, on-line computing system for the graphic demonstration of mathematical concepts and data which are important prerequisites for chemical calculations, thereby creating a "mathematical laboratory" for the simulation of chemical processes or systems students would otherwise not have seen.

Examples are the superimposition of sine waves to form a square wave

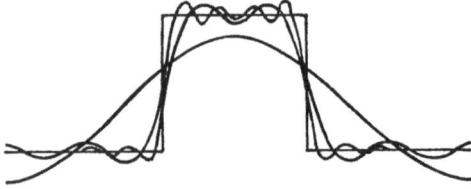

the angular dependence of the hydrogen-like atomic orbitals

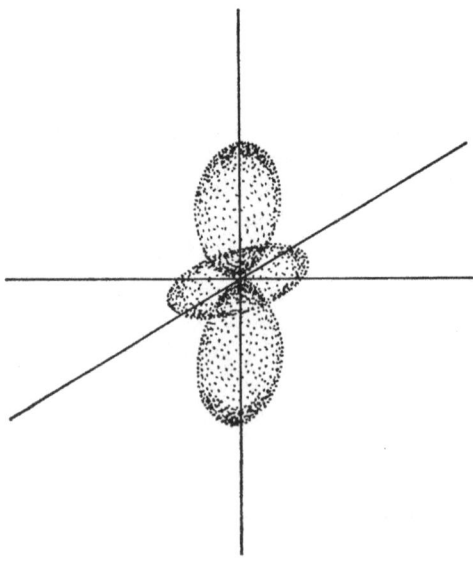

or the surface of functions obtained by solving the Schrödinger equation for a harmonic oscillator for a range of trial energies slightly above and below the first eigenvalue

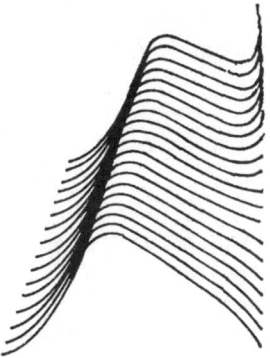

(Pictures are taken from [12].)

Computers may even be used to generate animated pictures of molecular vibrations (see for example, Mortensen and Penick [28].)

IV. Analog Computers

The analog computers presents itself as a kind of wallflower in the field of CAI. Although analog computers possess limited capabilities as compared with their 'big brothers', the digital computers, they nevertheless have tremendous potential as an aid in teaching the mathematics and dynamics of chemical reactions. A major advantage of analog simulations arises from the fact that analog computers are capable of solving several differential equations simultaneously, so that it is unneccessary to rely on steady-state solutions. This can, of course also be achieved by digital computers, but programming takes much more time and is irrelevant to the logic of the process.

Another advantage is that a large amount of interesting data can be obtained quite rapidly. As already mentioned, programming is simple, and the program (some connections on the board, certain positions of potentiometers) accurately reflects a chemical reaction. This swiftness of programming *and* of execution allows the student to simulate a large number of experiments in a laboratory period of few hours. Since the student can do his own programming, he not only learns to handle systems of differential equations but in addition appreciates that the computer is a powerful tool for solving problems of limited context.

Furthermore, output of data is swift and usually in graphical form which makes comprehension of complex phenomena much easier; to which is added the low cost and robustness of small instruments. Analog computers seem ideal for schools; unfortunately, they are rather despised by the computer world.

According to Corrin [9], and Griswold and Haugh [17], some of its applications are in showing:

The behavior of chemical reactions, especially the extend to which certain parameters (*e. g.* orders of reaction) are dependent on other parameters (*e. g.* rate constants).
The conditions under which a steady-state assumption may be valid.
The relationship between order and molecularity.
Wave functions of simple quantum-mechanical phenomena.

In short: everything apt to be described by differential equations. The book by Röpke and Riemann [34] gives over 40 examples — problems in inorganic chemistry (various kinds of reactions), organic chemistry (enzyme reactions), pharmaco-kinetics (drug adsorption), etc. Every problem is treated by graphic representation of the reaction, mathematical model, analog computer program, list of necessary elements, readout (character of curves), and a short discussion of results.

Griswold and Haugh [17], for example, show the student how to test the validity of a steady-state assumption. The student may choose inital values and certain other reaction parameters. The output will be the variation of concentration versus time, which, for a certain set of inital values, may look like this:

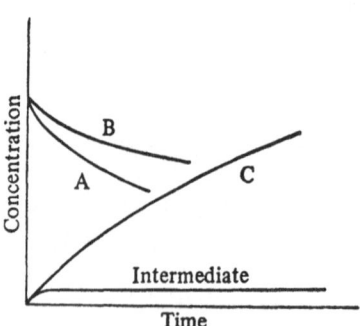

whereas with another set, the steady-state assumption proves to be invalid, as can be seen from the appropriate graph:

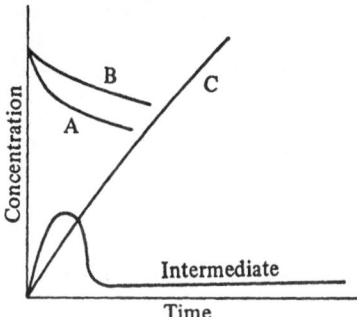

For other applications of analog computers see Knorre [23] and Tabbutt [44].

The reason why the anlog computer has been neglected as an instructional tool seems to be because it resembles a measuring instrument like an oscillograph (which it actually contains). Analog computers can't talk like digital computers are made to do, not always to the advantage of man-machine communication.

V. Literature

[1] Atkinson, R., Wilson, H.: Computer-assisted instruction. New York 1969.

[2] Avner, R. A., Tenczar, R.: The TUTOR Manual. CERL Report X-4. Urbana, Illinois: University of Illinois 1969.

[3] Bader, M.: Computer programs in undergraduate chemistry. J. Chem. Educ. 48, 3, 175 (1971).

[4] Baumgartner, W., Geist, W., Müller, C., Schmidt, B.: "CUS-Chemie"-Computer-unterstütztes Selbststudium Chemie. Naturw. Rundschau (Dez. 1972, in preparation).

[5] Bennik, F. D., Frye, C. H.: PLANIT Author's Guide. System Development Corporation, Santa Monica 1970.

[6] Bitzer, D. L., Blomme, R. W., Sherwood, B. A., Tenczar, P.: The PLATO System and Science Education. University of Illinois, CERL-Report X-17. Urbana, Illinois: Computerbased Education Research Laboratory (CERL) 1970.

[7] Burkhart, H.: STIDON — Ein Lehrprogramm mit Versuchssimulation zur Erkennung von Stoffen mittels chemischer Identifikationsmethoden. Paper, workshop on simulation, GPI-conference, April 5—8, Berlin 1972.

[8] Castleberry, S., Lagowski, J. J.: Individualized instruction using computer techniques. J. Chem. Educ. 47, 2, 91—96 (1970).

[9] Corrin, M. L.: Lecture demonstrations with an analog computer. J. Chem. Educ. 43, 579—581 (1966).

[10] Craig, N. C., Sherertz, D. D., Carlton, T. S., Ackermann, M. N.: Computer experiments some principles and examples. J. Chem. Educ. *48*, 5, 310—313 (1971).

[11] Dye, J. L., Nicely, V. A.: A general purpose curvefitting program for class and research use. J. Chem. Educ. *48*, 7, 445—448 (1971).

[12] Ewig, C. S., Geig, J. T., Harris, D. O.: An interactive on-line computing system as an instructional aid. J. Chem. Educ. *47*, 97—101 (1970).

[13] Feuerzeig, W., Swets, J. A.: Computer-aided instruction. Science *150*, 572—576 (1964).

[14] Gerard, R. W.: Computers and education. New York: McGraw-Hill 1967.

[15] Grandey, R.: An investigation of the use of computer-aided instruction in teaching students how to solve selected multistep general chemistry problems. CERL Report X-19, Urbana: University of Illinois 1970.

[16] Grandey, R.: The use of computers to aid instruction in beginning chemistry. J. Chem. Educ. *48*, 791—794 (1971).

[17] Griswold, R., Haugh J. F.: Analog computer simulation J. Chem. Educ. *45*, 576—580 (1968).

[18] Haefner, K.: Strategien des Lehrens I. Data Report *6*, 28—33 (1971).

[19] Haefner, K.: Strategien des Lehrens II. Data Report *7* (1972).

[20] Haefner, K., Ripota, P., Schramke, H.: Einführung in ICU/PLANIT. München: Siemens AG 1972.

[21] Holtzman, W. H. (ed.): Computer-assisted instruction, testing and guidance. New York: Harper and Row 1970.

[22] Coursewriter III: Version 3 Author's Guide. IBM, White Plains, New York, August 1971.

[23] Knorre, W. A.: Analog Computer in Biologie und Medizin. Jena: VEB Gustav Fischer Verlag 1971.

[24] Lagowski, J. J.: Computer-assisted instruction in chemistry. In: Holtzman, ed. etc., p. 283—298.

[25] Lambe, E. D.: Simulated laboratory in the natural sciences. In: Holtzman, ed. etc., p. 299—306.

[26] Lekan, H. A.: Index to computer-assisted instruction. New York: Third Edition Harcourt Brace Jovanovich, Inc. 1971.

[27] Lower, S. K.: Audio-tutorial and CAI aids. J. Chem. Educ. *47*, 2, 143—146 (1970).

[28] Mortensen, E. M., Penick, R. J.: Computer animation of molecular vibrations: Ethane. J. Chem. Educ. *47*, 102—104 (1970).

[29] Perone, S. P., Eagleston, J. F.: On-line digital computer applications in gas-chromatography — An undergraduate analytical experiment. J. Chem. Educ. *48*, 7, 438—442 (1971).

[30] Peterson, D. L., Fuller, M. E.: Physical chemistry students discover the computer. J. Chem. Educ. *48*, 314—316 (1971).

[31] Portugal, L. D., Minicozzi, W. P.: Computer generated display and manipulation of a general molecule. J. Chem. Educ. *48*, 790 (1971).

[32] Ripota, P.: Modell, Simulation, Spiel. Paper, workshop on simulation, GPI-conference, Berlin, April 5—8, 1972.

[33] Rodewald, L. B., Culp, G. H., Lagowski, J. J.: The use of computers in organic chemistry instruction. J. Chem. Educ. *47*, 2, 134—136 (1970).

[34] Röpke, H., Riemann, J.: Analogcomputer in Chemie und Biologie. Berlin-Heidelberg-New York: Springer 1969.

[35] Schefe, P., Wendler, D.: Simulation im naturwissenschaftlichen Unterricht. Paper, workshop on simulation, GPI-conference, Berlin, April 5—8, 1972.

[36] Lidia: Autorensprache für rechnergestützte Unterweisung. München: Siemens AG 1970.

37) A look at some of the CAI Programs relating to Chemistry at Simon Fraser University. Simon Fraser University, Department of Chemistry, Burnaby, Canada.

38) Computer assisted instruction at SFU. Simon Fraser University, Burnaby, Canada.

39) Simon, H.: Simulation im Rechnerunterstützten Unterricht. Paper, workshop on simulation, GPI-conference, Berlin, April 5—8, 1972.

40) Smith, S. G.: The use of computers in the teaching of organic chemistry. J. Chem. Educ. 47, 608—611 (1970).

41) Smith, S. G.: Computer-aided teaching of organic synthesis. J. Chem. Educ. 48, 727—729 (1971).

42) Soltzberg, L. J.: "Qualitative" computing in elementary chemical education. J. Chem. Educ. 48, 7, 449—452 (1971).

43) Suppes, P., Morningstar, M.: Computer assisted instruction. Science 166, 343—350 (1969).

44) Tabutt, F. D.: The use of analog computers for teaching chemistry. J. Chem. Educ. 44, 65—69 (1967).

45) Tansey, P. J., Unwin, D.: Simulation and gaming in education. London: Methuen Educational Ltd. 1969.

46) Wilkins, C. L., Klopfenstein, C. E.: Simulation of NMR-spectra computer as teaching devices. J. Chem. Educ. 43, 1, 10—13 (1966).

47) Yaney, N. D.: Computer system for individual home work: Keycard assembly, grading, and grading summation. J. Chem. Educ. 48, 4, 276—277 (1971).

Received August 21, 1972

SPRINGER-VERLAG
BERLIN·HEIDELBERG·NEW YORK

INTERNATIONAL COMPENDIUM OF NUMERICAL DATA PROJECTS

A Survey and Analysis

Produced by CODATA.
The Committee on Data
for Science and Technology
of the International Council
of Scientific Unions (ICSU)

XXIII, 295 pages. 1969
Cloth DM 48,—

The Compendium is the first comprehensive worldwide survey and analysis of centers which compile, evaluate, and publish numerical data for science and technology. The organization and scope of more than 150 data centers and projects in 26 countries are described in detail and their publications listed and reviewed. The centers and their output, which is vital to both science and industry, are arranged in the Compendium according to six broad property categories, as follows: nuclear; atomic and molecular, including spectroscopy; solid state (crystallographic, mineralogical, electrical, magnetic); thermodynamic, including transport and solution; chemical kinetics and other properties, including gas chromatographic and optical. Approx. 120 handbooks are listed, covering the above property categories and biology, earth sciences and analytical chemistry. Sources of internationally approved units, symbols, constants, and nomenclature are detailed.

Contents

National Data Programs and National Committees for CODATA.
Centers Covering a Number of Areas of Science.
Continuing Numerical Data Projects and their Publications.
New and Secondary Centers.
Handbooks and other Sources of Useful Tabular Data.
Physical Quantities, Units and Symbols, Basic Physical Constants, Nomenclature, Related Matters.
Author Index. Subject Index. Country Index. International Projects — International Unions Index.

HMO
Hückel Molecular Orbitals

From the reviews:

E. **Heilbronner**
and **P. A. Straub**

With 816 pages
DIN A 4
Loose Leaf. 1966
DM 92,—

"In 1961, when Streitwieser wrote Molecular Orbital Theory for Organic Chemists, he drew attention to the very rapid recent growth of interest in this field—seventy papers in the forties, 600 in the fifties, and a corresponding increase in the sixties. These π-electron molecular orbitals are usually represented as linear combinations of atomic orbitals (LCAO) with certain other approximations as introduced by Hückel. The enormous use of these Hückel MOs has now led to no less than three fullscale publications of tables of the relevant coefficients. The present volume, prepared by Prof. Heilbronner and Dr. Straub, is the latest attempt to provide the coefficients in the MOs and certain other dependent quantities in such a form as to be helpful to chemists who have no desire to make these calculations for themselves." (Nature)

Springer-Verlag
Berlin · Heidelberg · New York
München · London · Paris · Sydney · Tokio · Wien

In kritischen Übersichten werden in dieser Reihe Stand und Entwicklung aktueller chemischer Forschungsgebiete beschrieben. Sie wendet sich an alle Chemiker in Forschung und Industrie, die am Fortschritt ihrer Wissenschaft teilhaben wollen.

In der Regel werden nur Beiträge veröffentlicht, die ausdrücklich angefordert worden sind. Schriftleitung und Herausgeber sind aber für ergänzende Anregungen und Hinweise jederzeit dankbar. Manuskripte können in den ,,Fortschritten der chemischen Forschung" in Deutsch oder Englisch veröffentlicht werden.

Jeder Band der Reihe ist einzeln käuflich.

This series presents critical reviews of the present position and future trends in modern chemical research. It is addressed to all research and industrial chemists who wish to keep abreast of advances in their subject.

As a rule, contributions are specially commissioned. The editors and publishers will, however, always be pleased to receive suggestions and supplementary information. Papers are accepted for "Topics in Current Chemistry" in either German or English.

Any volume of the series may be purchased separately.